New Research in Obsessive-Compulsive Disorder and Major Depression

New Research in Obsessive-Compulsive Disorder and Major Depression

Special Issue Editors

Bruno Aouizerate
Emmanuel Haffen

MDPI • Basel • Beijing • Wuhan • Barcelona • Belgrade

MDPI

Special Issue Editors
Bruno Aouizerate
Pôle de Psychiatrie Géenérale et Universitaire,
Centre de référence régional des pathologies anxieuses et de la dépression,
Centre Expert Dépression Résistante FondaMental,
CH Charles Perrens, NutriNeuro, UMR INRA 1286,
Université de Bordeaux, 33076 Bordeaux, France
France

Emmanuel Haffen
University Hospital of Besançon
France

Editorial Office
MDPI
St. Alban-Anlage 66
4052 Basel, Switzerland

This is a reprint of articles from the Special Issue published online in the open access journal *Brain Sciences* (ISSN 2076-3425) in 2018 (available at: https://www.mdpi.com/journal/brainsci/special_issues/OCD_major_depression)

For citation purposes, cite each article independently as indicated on the article page online and as indicated below:

LastName, A.A.; LastName, B.B.; LastName, C.C. Article Title. *Journal Name* **Year**, *Article Number*, Page Range.

ISBN 978-3-03921-090-9 (Pbk)
ISBN 978-3-03921-091-6 (PDF)

Contents

About the Special Issue Editors

Bruno Aouizerate Bruno Aouizerate is professor of psychiatry at the University of Bordeaux (France). He is the head of the regional reference center for the assessment and management of anxiety and depressive disorders (Hospital Charles Perrens, Bordeaux) and an active member and coordinator of the French network of Expert Centers for Resistant Depression within the Fondation FondaMental. He also belongs to the nutrition and psychoneuroimmunology research team in the NutriNeuro laboratory (UMR INRA 1286, Bordeaux). He has coordinated or participated in several funded human research projects mainly focused on the study of pathophysiological determinants and therapeutic applications in major depression and obsessive-compulsive disorder. He is in charge of the axis "Mental disorders" of the FHU "Diagnosis, Prevention and Treatment of Neurological, Psychiatric, Metabolic and Sleep Disorders". Finally, he has authored more than 100 peer-reviewed publications.

Emmanuel Haffen (MD, PhD, University Hospital of Besançon) is professor of psychiatry at the University Hospital of Besançon. He leads the Laboratory of Clinical and Integrative Neurosciences of the University of Bourgogne Franche-Comté and the Center of Clinical Investigation of the University Hospital of Besançon (CIC 1431-INSERM). He coordinates the research networks of the Fondation FondaMental and is the president of AFPBN (French Association of Biological Psychiatry, member of the WFSBP).

brain sciences

MDPI

Editorial

New Research in Obsessive-Compulsive Disorder and Major Depression

Bruno Aouizerate [1,*] and Emmanuel Haffen [2,*]

1 Pôle de Psychiatrie Générale et Universitaire, Centre de référence régional des pathologies anxieuses et de la dépression, Centre Expert Dépression Résistante FondaMental, CH Charles Perrens, NutriNeuro, UMR INRA 1286, Université de Bordeaux, 33076 Bordeaux, France
2 Service de Psychiatrie, CIC-1431 INSERM, CHU de Besançon, laboratoire de Neurosciences, univ. Bourgogne-Franche-Comté, 25030 Créteil, France
* bruno.aouizerate@u-bordeaux.fr (B.A.); emmanuel.haffen@univ-fcomte.fr (E.H.)

Received: 13 June 2019; Accepted: 14 June 2019; Published: 17 June 2019

Major depression and obsessive-compulsive disorder (OCD) are among the most frequent psychiatric disorders in the general population. They are also frequently associated with one another in daily clinical practice. They generate severe emotional distress and marked impairment in general functioning. They are both characterized by a chronic and/or recurrent course, thereby leading to a profound deterioration of quality of life. Despite significant advances in pharmacological and psychological therapies, 20–30% of patients still respond unsuccessfully to standard medical treatment strategies for these disorders. In this context, one of the major outstanding research challenges is to better identify precise phenotypic profiles through the validation of innovative numerical tools, enabling the online recording and monitoring of the cognitive, emotional, and behavioral components underlying the expression of clinical phenotypes. This is nicely addressed in the article by Briffault et al. [1] from the perspective of opening new "symptoms networks" for the promotion of more homogenous diagnostic categories in psychiatry. In parallel, the development of relevant experimental paradigms is useful for the assessment of disturbances in sensory processes related to the olfactory sphere, which are classically observed in major depression, as part of the core dimensions/symptoms, such as anhedonia, primarily referring to a reduced feeling of pleasure from usually enjoyable activities. This kind of research receives particular attention from Rochet et al. [2] considering both anatomical and functional findings and relying on a translational approach ranging from laboratory animal models to humans for the determination of the brain regions that are more commonly implicated in olfaction, hedonic processes, and major depression. Additionally, there is a large body of literature supporting the implication of biological mechanisms, among which immune function with an abnormal inflammatory response is most often cited. This "inflammatory" theory, which was largely documented in the area of depression, was more recently extended to OCD, as in the article by Lamothe et al. [3] which examines cytokine profiles and white blood cell populations. The authors have also investigated relationships with streptococcal infections and the PANDAS generating a large constellation of neuropsychiatric conditions, especially including OCD that is mainly mediated by disruption in the basal ganglia, which are particularly vulnerable to antineuronal antibodies relying on stimulated autoimmune processes. Beyond the description of these pathophysiological determinants, the articles by Brunelin et al. [4] and Bennabi et al. [5] are respectively dedicated to therapeutic challenges in the utilization of the non-invasive brain stimulation technique tDCS in chronic forms of OCD and major depression that are unresponsive to the conventional medical treatments. Although data are still scarce, tDCS seems to represent a promising alternative for the management of major depression and OCD in substantially reducing the clinical severity with relatively good tolerability. However, these beneficial effects remain to be further confirmed in larger and controlled studies in order to alleviate the methodological concerns that could be raised in the already published trials. To conclude, this Special Issue will therefore provide precious information at the clinical, pathophysiological, and therapeutic levels that could be helpful to a wide readership, ranging from mental health professionals to basic/clinical researchers with a significant interest in increasing their knowledge in the field of major depression and OCD.

Conflicts of Interest: The authors declare no conflict of interest.

References

1. Briffault, X.; Morgiève, M.; Courtet, P. From e-Health to i-Health: Prospective Reflexions on the Use of Intelligent Systems in Mental Health Care. *Brain Sci.* **2018**, *8*, 98. [CrossRef] [PubMed]
2. Rochet, M.; El-Hage, W.; Richa, S.; Kazour, F.; Atanasova, B. Depression, Olfaction, and Quality of Life: A Mutual Relationship. *Brain Sci.* **2018**, *8*, 80. [CrossRef] [PubMed]
3. Lamothe, H.; Baleyte, J.; Smith, P.; Pelissolo, A.; Mallet, L. Individualized Immunological Data for Precise Classification of OCD Patients. *Brain Sci.* **2018**, *8*, 149. [CrossRef] [PubMed]
4. Brunelin, J.; Mondino, M.; Bation, R.; Palm, U.; Saoud, M.; Poulet, E. Transcranial Direct Current Stimulation for Obsessive-Compulsive Disorder: A Systematic Review. *Brain Sci.* **2018**, *8*, 37. [CrossRef] [PubMed]
5. Bennabi, D.; Haffen, E. Transcranial Direct Current Stimulation (tDCS): A Promising Treatment for Major Depressive Disorder? *Brain Sci.* **2018**, *8*, 81. [CrossRef] [PubMed]

brain
sciences

MDPI

Review

Transcranial Direct Current Stimulation for Obsessive-Compulsive Disorder: A Systematic Review

Jérôme Brunelin [1,2,3,4,*], Marine Mondino [1,2,3,4], Rémy Bation [1,2,3,4,5], Ulrich Palm [6], Mohamed Saoud [1,2,3,4,5] and Emmanuel Poulet [1,2,3,4,7]

1   INSERM, U1028, Lyon Neuroscience Research Center, PSY-R2 team, F-69000 Lyon, France;
    marine.mondino@ch-le-vinatier.fr (M.M.); remy.bation@chu-lyon.fr (R.B.);
    mohamed.saoud@chu-lyon.fr (M.S.); emmanuel.poulet@chu-lyon.fr (E.P.)
2   CNRS, UMR5292, Lyon Neuroscience Research Center, PSY-R2 Team, F-69000 Lyon, France
3   University Lyon, F-69000 Lyon, France
4   Centre Hospitalier le Vinatier, F-69678 Bron, France
5   Psychiatry Unit, Wertheimer Hospital, CHU Lyon, F-69500 Bron, France
6   Department of Psychiatry and Psychotherapy, Klinikum der Universität München,
    D-80336 Munich, Germany; Ulrich.Palm@med.uni-muenchen.de
7   Psychiatry Emergency Unit, Edouard Herriot Hospital, CHU, F-69000 Lyon, France
*   Correspondence: jerome.brunelin@ch-le-vinatier.fr; Tel.: +33-4-3791-5565

Received: 9 January 2018; Accepted: 23 February 2018; Published: 24 February 2018

Abstract: Despite the advances in psychopharmacology and established psychotherapeutic interventions, more than 40% of patients with obsessive-compulsive disorder (OCD) do not respond to conventional treatment approaches. Transcranial direct current stimulation (tDCS) has been recently proposed as a therapeutic tool to alleviate treatment-resistant symptoms in patients with OCD. The aim of this review was to provide a comprehensive overview of the current state of the art and future clinical applications of tDCS in patients with OCD. A literature search conducted on the PubMed database following PRISMA guidelines and completed by a manual search yielded 12 results: eight case reports, three open-label studies (with 5, 8, and 42 participants), and one randomized trial with two active conditions (12 patients). There was no sham-controlled study. A total of 77 patients received active tDCS with a large diversity of electrode montages mainly targeting the dorsolateral prefrontal cortex, the orbitofrontal cortex or the (pre-) supplementary motor area. Despite methodological limitations and the heterogeneity of stimulation parameters, tDCS appears to be a promising tool to decrease obsessive-compulsive symptoms as well as comorbid depression and anxiety in patients with treatment-resistant OCD. Further sham-controlled studies are needed to confirm these preliminary results.

Keywords: OCD; tDCS; brain stimulation; neuromodulation; obsession; compulsion

1. Introduction

Obsessive-compulsive disorder (OCD) is a frequent and debilitating psychiatric condition that occurs in 2–3% of the population [1]. Symptoms consist of unwanted intrusive thoughts and compulsive behaviours, leading to the inability to maintain social and occupational functioning [2].

Established treatments consist of a combination of psychopharmacology (especially selective serotonin reuptake inhibitor—SSRI) and psychotherapeutic interventions, such as cognitive behavioral therapy—CBT [3]. Despite augmentation strategies with other psychotropic drugs and advances in psychopharmacology [4], it is assumed that nearly 40% of patients do not show a sufficient response to conventional treatments [3]. Therefore, the development of new therapeutic approaches is warranted.

Among the recently developed therapeutic approaches, non-invasive brain stimulation techniques (NIBS), such as transcranial direct current stimulation (tDCS) hold promises to alleviate symptoms and improve cognitive functioning in various psychiatric conditions [5,6]. tDCS consists of applying a weak direct current (1–2 mA) between two electrodes placed on the scalp of a subject. Neurophysiological studies have reported that depending on the electrode polarity and current intensity, tDCS may increase cortical excitability in the vicinity of the anode whereas cathodal tDCS may decrease it [7]. The effects of tDCS are not restricted to the area beneath the electrodes and could reach a widespread network of cortical and subcortical regions that are connected to the targeted region [8]. The ability of tDCS to modulate a network is of particular interest since abnormal activity and connectivity within the orbitofronto-striato-pallido-thalamic network is described in patients with OCD. Indeed, imaging studies in patients with OCD showed abnormalities, which can be either hyper- or hypo activities, within numerous brain regions along a widespread network including the orbitofrontal cortex (OFC), the (pre-) supplementary motor area (SMA), the cingulate gyrus, the caudate, the thalamus, the right and left cerebellum, and the parietal cortex [9]. These abnormalities, which can be either trait- or state-dependent, have been revealed in resting conditions as well as by symptom provocation paradigms, depending on the studies. Moreover, it has been reported that some of these abnormalities were reverted after successful treatment [10,11]. It has thus been hypothesized that applying tDCS over these abnormal brain regions would lead to a decrease in obsessive-compulsive (OC) symptoms by modulating the underlying abnormal brain network. For instance, the use of anodal tDCS over the pre-SMA is based on imaging studies revealing an interaction between pre-SMA hypoactivity and deficient response inhibition with reciprocal striatal hyperactivity in patients with OCD [12]. The use of cathodal tDCS over the OFC is based on imaging studies reporting hyperactivity at rest and during symptom provocation paradigms of the OFC in patients with OCD [9,13]. Targeting the dorsolateral prefrontal cortex (DLPFC) is based on NIBS studies reporting beneficial clinical effects when stimulating this specific brain region in numerous psychiatric conditions [5,6], and on imaging studies reporting abnormalities in the cortico-striato-thalamo-cortical pathways, especially the 'DLPFC-caudate nucleus-thalamus' loop that is implicated in the pathophysiology of OCD [13]. This review aimed to provide a comprehensive overview of existing literature on the effects of tDCS applied as a therapeutic tool to reduce OC symptoms in patients with treatment-resistant OCD and to discuss future applications of tDCS in OCD.

2. Materials and Methods

Search Strategy

A systematic review was conducted following the recommendations of the PRISMA guidelines. A primary search on the PubMed database until December 2017 with the keywords (tDCS AND OCD) yielded 21 results. This primary search was completed by a manual search on articles cited by retrieved articles and on Google allowing for adding five articles (see Figure 1—PRISMA diagram).

The inclusion criteria were: (i) full length original articles published in English language in peer-reviewed journals, (ii) patients with OCD according to DSM or ICD-10 criteria, (iii) detailed description of the stimulation method, and (iv) the use of repeated sessions of tDCS. Among the 26 articles from the primary search, 11 articles were excluded for the following reasons: six were review articles not specifically dealing with tDCS in OCD, four did not concern OCD, and one offered a modelling of the electrical field induced by tDCS in patients with OCD. One of the articles issued by the manual search showed no data on OC symptoms [14], as well as a study investigating the clinical interest of transcranial alternating current stimulation (tACS) and not tDCS [15] were excluded from the qualitative analysis. Another article investigating the effect of a single session of tDCS (anode, cathode, sham) over the medial prefrontal cortex (PFC) on anxiety symptoms after exposure in 12 patients with treatment resistant OCD was also excluded [16].

A total of 12 articles was included in the qualitative analysis, nine from the primary search and three from the manual search [17–19].

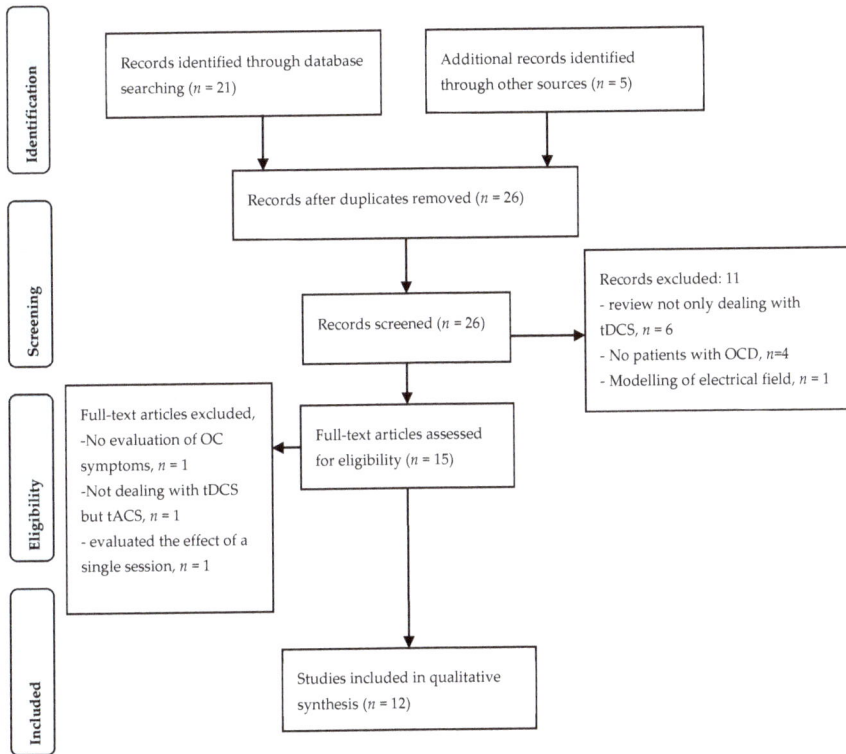

Figure 1. PRISMA flow diagram of selected studies in the qualitative analysis.

3. Results

Amongst the 12 included studies investigating the clinical effects of tDCS in patients with OCD, eight were case reports [17,19–25], three were open-label studies, including 5, 8, and 42 patients [18,26,27] and one was a randomized-controlled study including 12 patients with OCD [28]. Remarkably, none of the studies was sham-controlled (Table 1).

In the first case report, Volpato and colleagues [20] observed no significant effects of 10 sessions of tDCS (20 min, 2 mA) on OC symptoms when the cathode was placed over the left dorsolateral prefrontal cortex (DLPFC; over F3, according to the 10/20 international electroencephalography EEG system) and the anode extra-cephalically (on the neck). Interestingly, the authors reported a significant decrease of depression and anxiety symptoms. Other studies have tried different electrode montages and have shown beneficial outcomes on OC symptoms (see Figure 2 for an illustration of the electrode montages). Namely, two studies targeted the left DLPFC by placing the anode over the left DLPFC (F3) and the cathode either over the right DLPFC (F4) [25] or the right orbitofrontal cortex (OFC) /supraorbital area (Fp2) [18]. Three studies used an electrode montage positioning the cathode over the left OFC (Fp2) and the anode over the occipital region (O2) or the cerebellum [19,22,26]. One study targeted the right OFC (Fp2) with the cathode and the left parieto-temporo-occipital region with the anode (midway between P1, C3, and T7) [27]. Finally, five studies targeted the pre-supplementary motor area (SMA). Among them, two placed the anode over the pre-SMA (Fz/FCz) and the cathode over the right orbitofrontal cortex (Fp2) [17,21], one placed the cathode over the pre-SMA and the anode extra-cephalically over the right deltoid [24], and two compared two different montages with either the anode or the cathode over the pre-SMA and the other electrode extra cephalically over the right deltoid [23,28]. In the included studies, different sizes of electrodes were used: 25 cm^2 [19,23,24,28], 35 cm^2 [18,20–22,25,26] and 5.5 cm^2 [27]. The intensity of stimulation was set at 2 mA in all of the tDCS studies (2–3 mA in [27]) and tDCS duration varied from 20 min [17–20,22,23,26,28] to 30 min [24,25,27]. The number of tDCS sessions also varied; most of the studies delivered 10 [19,20,22,23,26,28] or 20 sessions [17,21,24,25] and one study delivered 15 sessions [18]. tDCS sessions were delivered daily [18,19,23,24,27,28] or twice daily [17,21,22,25,26]. All of the studies used the Yale–Brown Obsessive and Compulsive Scale score (Y-BOCS) [29] to assess OC symptoms.

In summary, a total of 77 patients with OCD received active tDCS with different electrode montages. Most of the studies reported a significant effect of tDCS on OC symptoms, more specifically, a decrease of the YBOCS score. Several studies also reported beneficial effects of tDCS on other symptoms that are often observed in patients with OCD, such as depression and anxiety [18–20,24,25].

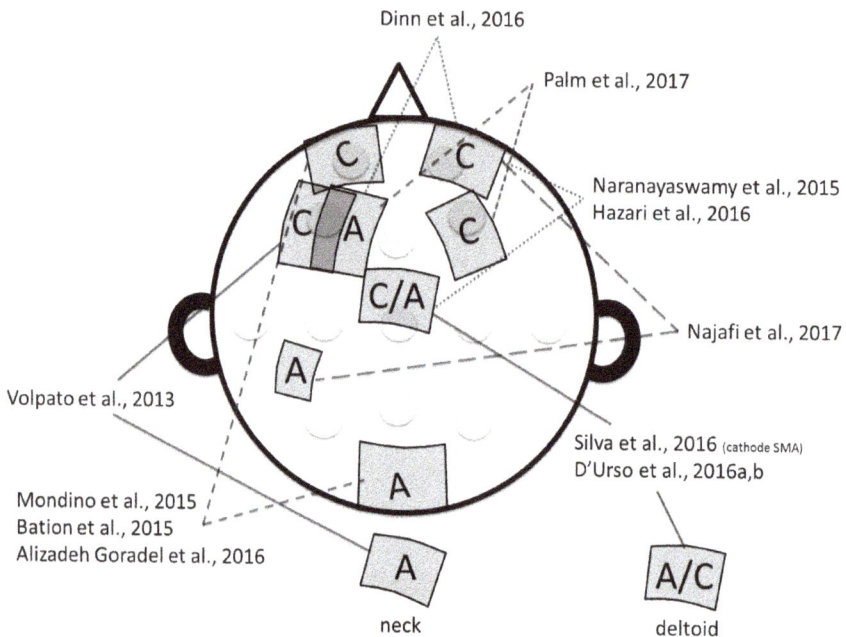

Figure 2. Illustration of the diversity in electrodes montage observed in transcranial direct current stimulation (tDCS) studies aiming to alleviate obsessive-compulsive symptoms in patients with treatment-resistant obsessive-compulsive disorder. A: Anode; C: Cathode. Hazari et al., 2016 [17]; Dinn et al., 2016 [18]; Alizadeh Goradel et al., 2016 [19]; Volpato et al., 2013 [20]; Narayanaswamy et al., 2015 [21]; Mondino et al., 2015 [22]; D'Urso et al., 2016a [23]; Silva et al., 2016 [24]; Palm et al., 2017 [25]; Bation et al., 2015 [26]; Najafi et al., 2017 [27]; D'Urso et al., 2016b [28].

Table 1. Main findings of studies investigating the clinical interest of transcranial direct current stimulation (tDCS) to decrease symptoms in patients with obsessive-compulsive disorder (OCD).

Articles	N	Patients Characteristics	Target	Intensity Electrode Size	Duration and Number of Sessions	Main Results
Volpato et al., 2013 [20]	1	Age: 35, male Type*: 2,3,4 Previous TTT: SSRI, SNRI, CBT	Anode: posterior neck-base Cathode: left DLPFC	2 mA, 35 cm²	20 min, 10 sessions (1/day)	No effect on OC symptoms. Depression score decreased (−34% HDRS); anxiety score decreased (−17%).
Mondino et al., 2015 [22]	1	Age: 52, female Type*: 3,4 Previous TTT: tricyclic, SSRI, SNRI, AP, Lithium, CBT	Anode: right cerebello-occipital (100 cm²) Cathode: left OFC	2 mA, 35 cm²	20 min, 10 sessions (2/day; 2 h between 2 sessions)	YBOCS score decreased (−26%)
Hazari et al., 2016 [17]	1	Age: 24, male Type*: 1,2 Previous TTT: SSRI, ECT	Anode: SMA Cathode: right OFC	2 mA, ND	20 min, 20 sessions (2/day, at least 3 h between 2 sessions)	YBOCS decreased (−80%) during 7 months
D'Urso et al., 2016 [23]	1	Age: 33, female Type*: 3 Previous TTT: SSRI, BZD, tricyclic, CBT	Anode: Pre-SMA Cathode: right deltoid And then, Reverse montage	2 mA, 25 cm²	20 min, 10 sessions (1/day)	Worsening of symptoms after anodal tDCS. YBOCS score decreased (−30%) after cathodal tDCS.
Alizadeh Goradel et al., 2016 [19]	1	Age: 23, female Type*: 1 Previous TTT: SSRI	Anode: right occipital Cathode: left OFC	2 mA, 25 cm²	20 min, 10 sessions (1/day)	YBOCS score decreased (−64%); Depression score decreased (−87%); −100% anxiety
Palm et al., 2017 [25]	1	Age: 31, male Type*: 1,3 Previous TTT: tricyclic, SSRI, AP, CBT	Anode: left DLPFC Cathode: Right DLPFC	2 mA, 35 cm²	30 min, 20 sessions (2/day, 3 h between 2 sessions)	Combined with Sertraline, YBOCS score (−22%), depression (−10%) and anxiety (−21%) decreased
Narayanaswamy et al., 2015 [21]	2	Age: 39, female Type*: 1 Previous TTT: SSRI, exposure	Anode: left pre-SMA Cathode: right OFC	2 mA, 35 cm²	20 min, 20 sessions (2/day, at least 3 h between 2 sessions)	Patient 1: YBOCS score decreased (−40%), −52% at day 17
		Age: 24, male Type*: 1 Previous TTT: tricyclic, SSRI				Patient 2: YBOCS score decreased (−46.7%)
Silva et al., 2016 [24]	2	Age: 37, male Type*: 2 Previous TTT: tricyclic, SSRI, CBT	Anode: right deltoid Cathode: bilateral SMA	2 mA, 25 cm²	30 min, 20 sessions (1/day)	Patient 1: no effect at Week 4, YBOCS score decreased at week 12 (−18%). No changes in anxiety nor depression
		Age: 31, male Type*: 1,3 Previous TTT: tricyclic, SSRI				Patient 2: YBOCS score decreased, (−17%) at Week 4; −55% at week 12). 50% improvement in anxiety and depression

Table 1. *Cont.*

Dinn et al., 2016 [18]	5	Age: 40.4 (8.4), 4 females, 1 male Type*: ND Previous TTT: SSRI, SNRI, AP	Anode: left DLPFC Cathode: right OFC	2 mA, 35 cm²	20 min, 15 sessions (1/day)	Open Label Study OC symptoms decreased (−23%); depression decreased (−30%)
Bation et al., 2015 [26]	8	Age: 44.2 (13.8), 6 females, 2 males Type*: 1 (n = 5), 3 (n = 3) Previous TTT: tricyclic, 3 SSRI, AP, CBT	Anode: right cerebellum Cathode: left OFC	2 mA, 35 cm²	20 min, 10 sessions (2/day, at least 3 h between 2 sessions)	Open Label Study YBOCS score decreased (−24.6%)
D'Urso et al., 2016 [29]	12	Age: 39.0 (13.1), 7 females, 5 males Type*: 1 (n = 4), 2 (n = 2), 3 (n = 6) Previous TTT: at least SSRI, CBT	Anode midline pre SMA Cathode: right deltoid (n = 6) OR reverse montage	2 mA, 25 cm²	20 min, 10 sessions (1/day)	RCT—10 patients completed the study Cathodal tDCS was significantly more effective than anodal tDCS. In cathodal arm, YBOCS score decreased (−17.5%) after 10 sessions, −20.1% after 20 sessions
Najafi et al., 2017 [27]	42	Age: 29.1 (10.1), 23 females, 19 males Type*: ND Previous TTT: at least 2 SSRI, CBT	Anode: parieto-temporo-occipital areas Cathode: right OFC	2–3 mA, 5.5 cm²	30 min, 15 sessions (1/day)	Open Label Study YBOCS score decreased (−63.4%) Maintenance of the effect at 3 months follow up (−77.6%)

DLPFC: dorsolateral prefrontal cortex; HDRS: Hamilton Depression Rating Scale; ND: Not Done; OFC: orbitofrontal cortex; (pre) SMA: (pre) supplementary motor area; Y-BOCS: Yale-Brown Obsessive Compulsive Scale. Age: mean (standard deviation) years; TTT: treatment; AP: antipsychotic, SSRI: selective serotonin reuptake inhibitor, SNRI: serotonin–norepinephrine reuptake inhibitor, CBT: cognitive behavioural therapy. Type* 1 = obsessions and checking, 2 = symmetry and ordering, 3 = cleanliness and washing, 4 = hoarding according to Leckman et al. 1997 [30].

4. Discussion

We reviewed here studies investigating the clinical effects of tDCS in patients with treatment-resistant OCD. Overall, our review included 12 studies, corresponding with a total sample of 77 patients with OCD. Results indicated that applying tDCS might show promising results to reduce OC symptoms. Little is known regarding the duration of this effect since it has not been systematically investigated. Two studies reported that the beneficial effects were still observed at a three-month [27] or seven-month follow-up [17]. In addition, it is interesting to note that some of the included studies also reported beneficial effects of tDCS on depression and anxiety that are common comorbid symptoms in patients with OCD. In line with this, a recent crossover study has investigated the effect of a single session of tDCS on obsession-induced anxiety after symptom provocation in patients with OCD. They reported a significant decrease in the severity of the obsession-induced anxiety following tDCS applied with the cathode over the medial PFC as compared with tDCS applied with the anode over the medial PFC and sham tDCS [16]. One may hypothesize that anxiety, depression, and OCD share abnormalities within brain networks that are targeted by cortical stimulation. However, the findings of beneficial effects of tDCS in OCD should be interpreted with caution and some methodological considerations should be noted.

First, none of the studies that are included in the present review was sham-controlled. One randomized study used a parallel arm design, but compared two active conditions [28]. To the best of our knowledge, only one randomized sham-controlled trial was conducted in OCD patients. This study included 20 patients with OCD and reported that active tDCS (2 mA, 20 min, 15 sessions) applied with the anode over the right DLPFC and the cathode over the left DLPFC improved decision-making abilities as compared to sham tDCS [14]. The authors thus showed that tDCS could have a pro-cognitive effect in patients with OCD, as reported in other psychiatric conditions [5]. However, this study did not provide direct clinical assessment of OC symptoms. Further studies are needed to determine the real effect of repeated sessions of active tDCS on OC symptoms, by comparing with sham. Indeed, previous sham-controlled studies have reported a large sham effect in patients with treatment-resistant OCD receiving repeated sessions of NIBS [11,31].

Second, most of the studies included in our review were case reports and only two studies included more than 10 patients [27,28]. Interpretation of results is thus limited by small sample size. Furthermore, tDCS parameters were highly heterogeneous across studies, in terms of electrode montage (see Figure 2), number of tDCS sessions, tDCS duration, and interval between sessions (from 2 h to 1 day). For instance, regarding electrode positioning, some studies targeted the DLPFC with the anode placed over the left DLPFC (F3) and the cathode over the right OFC [18], or the contralateral DLPFC [25]. Another study placed the cathode over the left DLPFC (F3) and the anode over the neck [20]. Other studies have proposed to target the left OFC (Fp1) or the right OFC (FP2) with the cathode combined with the anode over the right occipital cortex [19,22], the right cerebellum [26] or the temporo-parieto-occipital region [27]. These montages were based on neuroimaging studies showing hyperactivity within the left OFC and hypoactivity within the cerebellum in patients with OCD [9]. The pre-SMA was also commonly targeted in the reviewed studies either with the anode [17,21,23,28] or with the cathode [23,24,28]. In a randomized controlled trial comparing both montages (anode over the pre-SMA or cathode over the pre-SMA), D'Urso and colleagues suggested a better effect of the cathodal-tDCS montage on OC symptoms [28]. Nevertheless, based on these findings, it seems difficult to conclude regarding the optimal tDCS montage to adopt in order to alleviate symptoms in patients with OCD. However, in a computer head modelling study, Senço and colleagues reported interesting findings that may help us identifying the optimal electrode positioning [32]. More precisely, they found that the best theoretical montage to target the neurocircuitry involved in OCD would be with the cathode over the pre-SMA with an extra-cephalic anode, as done in D'Urso and colleagues' study [28].

Regarding the number and duration of tDCS sessions, the choice of delivering 10 to 20 sessions of 20 to 30 min has been mostly extrapolated from the data obtained in studies investigating the clinical effects of tDCS in patients with depression. However, it is not clearly established that increasing the duration and number of sessions leads to a better and longer clinical effect. The interval between consecutive tDCS sessions should also be considered. Indeed, some studies have shown that the inhibitory effects of a session of cathodal tDCS on motor corticospinal excitability were increased if a second tDCS session was performed during the after-effects of the first and were initially reduced and then re-established if the second tDCS session was performed 3 or 24 h after the first one [33]. Furthermore, it was reported in another study that the excitatory effects of anodal tDCS on motor corticospinal excitability were reduced, but prolonged when a second tDCS session was applied during the after-effects of the first (from 0 to 20 min after) but entirely abolished when the second tDCS session was applied 3 or 24 h after the first [34].

It is important to mention that most of the patients included in the reviewed studies were treated with different medication (in terms of duration and molecules) when they received tDCS. Most of the patients were treated with SSRIs, but in some studies, they were also treated with other medications, such as serotonin-norepinephrine reuptake inhibitors (SNRI), mood stabilizers or antipsychotics. The concomitant use of medication may influence the effects of tDCS [35]. For instance, studies investigating the effects of tDCS on motor corticospinal excitability have reported that both acute and chronic administration of the SSRI (citalopram) increased and prolonged the excitatory effects that are induced by anodal tDCS and reversed the inhibitory effects of cathodal tDCS into facilitation [36,37]. Furthermore, in a randomized-controlled trial in patients with major depressive disorder, Brunoni et al. have reported that combining tDCS with SSRI (sertraline hydrochloride) induced beneficial clinical improvements that were superior to each treatment taken separately (tDCS only or sertraline only) or sham [38]. Thus, future studies should take into account the concomitant use of medication when investigating the effects of tDCS on OC symptoms.

Clinical characteristics of patients should be taken into account when discussing the role of tDCS in the OCD treatment. For instance, the level of resistance was highly heterogeneous in the reviewed studies; some patients were resistant to several months of combination between SSRI and CBT (e.g., [18,22–26]), some others received ECT [17]. In addition, the subtypes of OCD (obsessions and checking; symmetry and ordering; cleanliness and washing; and, hoarding) [30] were also heterogeneous across studies and might be an important factor to report in future studies. These differences may also account to explain discrepancies observed between studies in term of symptoms improvement (from no effect on OC symptoms [20] to 80% decrease on YBOCS score [17]).

Another limitation is the brain state dependency that may have an impact on the tDCS clinical effect and should also be controlled in future studies. For instance, in the study reporting the largest beneficial effect of tDCS on OC symptoms, patients were not at rest during the stimulation session as in other studies but were required to listen to music and watch movies during the 30-min session duration [27]. In the same way, a single session of cathodal tDCS has been shown to have a beneficial effect when applied during exposure to anxiety [16]. Future studies should investigate the clinical effect of repeated sessions of tDCS when stimulation is applied during exposure to anxiety as compared to with being at rest.

Finally, up to now, no study has investigated the brain correlates of the symptom improvement following tDCS administration in OCD patients. Investigating biological effects of tDCS in patients with OCD will provide a better understanding of the pathophysiology of OCD (as done with other therapeutics, see [10,11]) and of tDCS' mechanisms of action. It could be speculated for example that hyperactive cortico-striatal pathways observed in patients with OCD may be down-regulated by either inhibitory stimulation of the OFC or SMA (with the cathode) or by excitatory stimulation (enhancement, with the anode) of the DLPFC. Neuroimaging or electrophysiological investigations are also needed to steer the parameter optimization. In this way, another transcranial electrical stimulation approach has been proposed recently by Klimke and colleagues [15]. In an open-label study including seven patients with OCD, the authors reported the clinical interest of transcranial alternating current stimulation (tACS) applied at gamma frequency (40 Hz). They observed that gamma-tACS applied in a bilateral fronto-temporal montage decreased OC symptoms by 52%, measured by the YBOCS. This novel protocol appears interesting and future studies are needed to further explore the effects of gamma-tACS in patients with OCD. Besides the optimization of stimulation parameters, further studies are also needed to determine the clinical and biological predictors of response, as done in studies on depression [39].

5. Conclusions

To conclude, only a few studies investigated the effects of tDCS in OCD, but they showed promising results, with some of them reporting a decrease >35% in YBOCS scores. This effect can be considered as clinically meaningful since the current definition of treatment response is at least a 35% reduction of Y-BOCS score [40]. However, these results are preliminary and further sham-controlled studies are needed to define the role of tDCS in the treatment of OCD and to determine the optimal stimulation parameters to deliver in this indication and subtypes of OCD. To date, regarding the high heterogeneity among studies in terms of the characteristics of patients (e.g., subtypes of OC symptoms, concomitant medication, age) and tDCS parameters (e.g., electrode montage, symptoms provocation paradigm during tDCS), it is difficult to draw a clear conclusion on the efficacy of tDCS in this indication and to propose guidelines for future investigations. Interestingly, based on these preliminary positive reports, randomized clinical trials have been initiated and are now recruiting participants around the world (as seen in clinical trials website: NCT 02407288, 02743715, 03304600). Results from these studies are expected before any conclusion on the relevance of tDCS in patients with OCD can be made.

Acknowledgments: The authors thank the "Conseil Scientific de la recherche du Centre Hospitalier le Vinatier" for financial support.

Author Contributions: J.B. and M.M. managed the literature search and assisted with the data collection. J.B., M.M. and R.B. wrote the first draft of manuscript. E.P. supervised searches. E.P., M.S. and U.P. critically revised the manuscript. All authors contributed to and have approved the final manuscript.

Conflicts of Interest: U.P. received speaker's honorarium from NeuroCare Group Munich, and has a private practice with NeuroCare Group, Munich. Other authors declare to not have conflict of interest.

References

1. Ruscio, A.M.; Stein, D.J.; Chiu, W.T.; Kessler, R.C. The epidemiology of obsessive-compulsive disorder in the National Comorbidity Survey Replication. *Mol. Psychiatry* **2010**, *15*, 53–63. [CrossRef] [PubMed]
2. Fontenelle, I.S.; Fontenelle, L.F.; Borges, M.C.; Prazeres, A.M.; Rangé, B.P.; Mendlowicz, M.V.; Versiani, M. Quality of life and symptom dimensions of patients with obsessive-compulsive disorder. *Psychiatry Res.* **2010**, *179*, 198–203. [CrossRef] [PubMed]
3. Fineberg, N.A.; Reghunandanan, S.; Simpson, H.B.; Phillips, K.A.; Richter, M.A.; Matthews, K.; Stein, D.J.; Sareen, J.; Brown, A.; Sookman, D. Accreditation Task Force of The Canadian Institute for Obsessive Compulsive Disorders Obsessive-compulsive disorder (OCD): Practical strategies for pharmacological and somatic treatment in adults. *Psychiatry Res.* **2015**, *227*, 114–125. [CrossRef] [PubMed]
4. Pallanti, S.; Quercioli, L. Treatment-refractory obsessive-compulsive disorder: Methodological issues, operational definitions and therapeutic lines. *Prog. Neuropsychopharmacol. Biol. Psychiatry* **2006**, *30*, 400–412. [CrossRef] [PubMed]
5. Mondino, M.; Bennabi, D.; Poulet, E.; Galvao, F.; Brunelin, J.; Haffen, E. Can transcranial direct current stimulation (tDCS) alleviate symptoms and improve cognition in psychiatric disorders? *World J. Biol. Psychiatry* **2014**, *15*, 261–275. [CrossRef] [PubMed]
6. Lefaucheur, J.-P.; Antal, A.; Ayache, S.S.; Benninger, D.H.; Brunelin, J.; Cogiamanian, F.; Cotelli, M.; de Ridder, D.; Ferrucci, R.; Langguth, B.; et al. Evidence-based guidelines on the therapeutic use of transcranial direct current stimulation (tDCS). *Clin. Neurophysiol.* **2017**, *128*, 56–92. [CrossRef] [PubMed]
7. Nitsche, M.A.; Paulus, W. Excitability changes induced in the human motor cortex by weak transcranial direct current stimulation. *J. Physiol.* **2000**, *527*, 633–639. [CrossRef] [PubMed]
8. Keeser, D.; Meindl, T.; Bor, J.; Palm, U.; Pogarell, O.; Mulert, C.; Brunelin, J.; Moller, H.-J.; Reiser, M.; Padberg, F. Prefrontal Transcranial Direct Current Stimulation Changes Connectivity of Resting-State Networks during fMRI. *J. Neurosci.* **2011**, *31*, 15284–15293. [CrossRef] [PubMed]

9. Hou, J.; Wu, W.; Lin, Y.; Wang, J.; Zhou, D.; Guo, J.; Gu, S.; He, M.; Ahmed, S.; Hu, J.; et al. Localization of cerebral functional deficits in patients with obsessive-compulsive disorder: A resting-state fMRI study. *J. Affect. Disord.* **2012**, *138*, 313–321. [CrossRef] [PubMed]

10. Van der Straten, A.L.; Denys, D.; van Wingen, G. Impact of treatment on resting cerebral blood flow and metabolism in obsessive compulsive disorder: A meta-analysis. *Sci. Rep.* **2017**, *7*, 17464. [CrossRef] [PubMed]

11. Nauczyciel, C.; le Jeune, F.; Naudet, F.; Douabin, S.; Esquevin, A.; Vérin, M.; Dondaine, T.; Robert, G.; Drapier, D.; Millet, B. Repetitive transcranial magnetic stimulation over the orbitofrontal cortex for obsessive-compulsive disorder: A double-blind, crossover study. *Transl. Psychiatry* **2014**, *4*, e436. [CrossRef] [PubMed]

12. Rauch, S.L.; Jenike, M.A.; Alpert, N.M.; Baer, L.; Breiter, H.C.; Fischman, A.J. Regional cerebral blood flow measured during symptom provocation in obsessive-compulsive disorder using oxygen 15-labeled carbon dioxide and positron emission tomography. *Arch. Gen. Psychiatry* **1994**, *51*, 62–70. [CrossRef] [PubMed]

13. Milad, M.R.; Rauch, S.L. Obsessive-compulsive disorder: Beyond segregated cortico-striatal pathways. *Trends Cogn. Sci.* **2012**, *16*, 43–51. [CrossRef] [PubMed]

14. Yekta, M.; Rostami, R.; Fayyaz, E. Transcranial Direct Current Stimulation of Dorsolateral Prefrontal Cortex in Patients with Obsessive Compulsive Disorder to Improve Decision Making and Reduce Obsession Symptoms. *Pract. Clin. Psychol.* **2015**, *3*, 185–194.

15. Klimke, A.; Nitsche, M.A.; Maurer, K.; Voss, U. Case Report: Successful Treatment of Therapy-Resistant OCD with Application of Transcranial Alternating Current Stimulation (tACS). *Brain Stimul.* **2016**, *9*, 463–465. [CrossRef] [PubMed]

16. Todder, D.; Gershi, A.; Perry, Z.; Kaplan, Z.; Levine, J.; Avirame, K. Immediate Effects of Transcranial Direct Current Stimulation on Obsession-Induced Anxiety in Refractory Obsessive-Compulsive Disorder: A Pilot Study. *J. ECT* **2017**. [CrossRef] [PubMed]

17. Hazari, N.; Narayanaswamy, J.C.; Chhabra, H.; Bose, A.; Venkatasubramanian, G.; Reddy, Y.C.J. Response to Transcranial Direct Current Stimulation in a Case of Episodic Obsessive Compulsive Disorder. *J. ECT* **2016**, *32*, 144–146. [CrossRef] [PubMed]

18. Dinn, W.M.; Aycicegi-Dinn, A.; Göral, F.; Karamursel, S.; Yildirim, E.A.; Hacioglu-Yildirim, M.; Gansler, D.A.; Doruk, D.; Fregni, F. Treatment-resistant obsessive-compulsive disorder: Insights from an open trial of transcranial direct current stimulation (tDCS) to design a RCT. *Neurol. Psychiatry Brain Res.* **2016**, *22*, 146–154. [CrossRef]

19. Alizadeh Goradel, J.; Pouresmali, A.; Mowlaie, M.; Sadeghi Movahed, F. The Effects of Transcranial Direct Current Stimulation on Obsession-compulsion, Anxiety, and Depression of a Patient Suffering from Obsessive-compulsive Disorder. *Pract. Clin. Psychol.* **2016**, *4*, 75–80. [CrossRef]

20. Volpato, C.; Piccione, F.; Cavinato, M.; Duzzi, D.; Schiff, S.; Foscolo, L.; Venneri, A. Modulation of affective symptoms and resting state activity by brain stimulation in a treatment-resistant case of obsessive-compulsive disorder. *Neurocase* **2013**, *19*, 360–370. [CrossRef] [PubMed]

21. Narayanaswamy, J.C.; Jose, D.; Chhabra, H.; Agarwal, S.M.; Shrinivasa, B.; Hegde, A.; Bose, A.; Kalmady, S.V.; Venkatasubramanian, G.; Reddy, Y.C.J. Successful Application of Add-on Transcranial Direct Current Stimulation (tDCS) for Treatment of SSRI Resistant OCD. *Brain Stimul.* **2015**, *8*, 655–657. [CrossRef] [PubMed]

22. Mondino, M.; Haesebaert, F.; Poulet, E.; Saoud, M.; Brunelin, J. Efficacy of Cathodal Transcranial Direct Current Stimulation Over the Left Orbitofrontal Cortex in a Patient With Treatment-Resistant Obsessive-Compulsive Disorder. *J. ECT* **2015**, *31*, 271–272. [CrossRef] [PubMed]

23. D'Urso, G.; Brunoni, A.R.; Anastasia, A.; Micillo, M.; de Bartolomeis, A.; Mantovani, A. Polarity-dependent effects of transcranial direct current stimulation in obsessive-compulsive disorder. *Neurocase* **2016**, *22*, 60–64. [CrossRef] [PubMed]

24. Silva, R.M.; Brunoni, A.R.; Miguel, E.C.; Shavitt, R.G. Transcranial direct current stimulation for treatment-resistant obsessive-compulsive disorder: report on two cases and proposal for a randomized, sham-controlled trial. *Sao Paulo Med. J.* **2016**, *134*, 446–450. [CrossRef] [PubMed]

25. Palm, U.; Leitner, B.; Kirsch, B.; Behler, N.; Kumpf, U.; Wulf, L.; Padberg, F.; Hasan, A. Prefrontal tDCS and sertraline in obsessive compulsive disorder: A case report and review of the literature. *Neurocase* **2017**, *23*, 173–177. [CrossRef] [PubMed]

26. Bation, R.; Poulet, E.; Haesebaert, F.; Saoud, M.; Brunelin, J. Transcranial direct current stimulation in treatment-resistant obsessive-compulsive disorder: An open-label pilot study. *Prog. Neuropsychopharmacol. Biol. Psychiatry* **2016**, *65*, 153–157. [CrossRef] [PubMed]

27. Najafi, K.; Fakour, Y.; Zarrabi, H.; Heidarzadeh, A.; Khalkhali, M.; Yeganeh, T.; Farahi, H.; Rostamkhani, M.; Najafi, T.; Shabafroz, S.; et al. Efficacy of Transcranial Direct Current Stimulation in the Treatment: Resistant Patients who Suffer from Severe Obsessive-compulsive Disorder. *Indian J. Psychol. Med.* **2017**, *39*, 573–578. [CrossRef] [PubMed]

28. D'Urso, G.; Brunoni, A.R.; Mazzaferro, M.P.; Anastasia, A.; de Bartolomeis, A.; Mantovani, A. Transcranial direct current stimulation for obsessive-compulsive disorder: A randomized, controlled, partial crossover trial. *Depress. Anxiety* **2016**, *33*, 1132–1140. [CrossRef] [PubMed]

29. Goodman, W.K.; Price, L.H.; Rasmussen, S.A.; Mazure, C.; Delgado, P.; Heninger, G.R.; Charney, D.S. The Yale-Brown Obsessive Compulsive Scale. II. Validity. *Arch. Gen. Psychiatry* **1989**, *46*, 1012–1016. [CrossRef] [PubMed]

30. Leckman, J.F.; Grice, D.E.; Boardman, J.; Zhang, H.; Vitale, A.; Bondi, C.; Alsobrook, J.; Peterson, B.S.; Cohen, D.J.; Rasmussen, S.A.; et al. Symptoms of obsessive-compulsive disorder. *Am. J. Psychiatry* **1997**, *154*, 911–917. [PubMed]

31. Mansur, C.G.; Myczkowki, M.L.; de Barros Cabral, S.; Sartorelli, M.D.C.B.; Bellini, B.B.; Dias, A.M.; Bernik, M.A.; Marcolin, M.A. Placebo effect after prefrontal magnetic stimulation in the treatment of resistant obsessive-compulsive disorder: A randomized controlled trial. *Int. J. Neuropsychopharmacol.* **2011**, *14*, 1389–1397. [CrossRef] [PubMed]

32. Senço, N.M.; Huang, Y.; D'Urso, G.; Parra, L.C.; Bikson, M.; Mantovani, A.; Shavitt, R.G.; Hoexter, M.Q.; Miguel, E.C.; Brunoni, A.R. Transcranial direct current stimulation in obsessive-compulsive disorder: Emerging clinical evidence and considerations for optimal montage of electrodes. *Expert Rev. Med. Devices* **2015**, *12*, 381–391. [CrossRef] [PubMed]

33. Monte-Silva, K.; Kuo, M.F.; Liebetanz, D.; Paulus, W.; Nitsche, M.A. Shaping the Optimal Repetition Interval for Cathodal Transcranial Direct Current Stimulation (tDCS). *J. Neurophysiol.* **2010**, *103*, 1735–1740. [CrossRef] [PubMed]

34. Monte-Silva, K.; Kuo, M.-F.; Hessenthaler, S.; Fresnoza, S.; Liebetanz, D.; Paulus, W.; Nitsche, M.A. Induction of Late LTP-Like Plasticity in the Human Motor Cortex by Repeated Non-Invasive Brain Stimulation. *Brain Stimul.* **2013**, *6*, 424–432. [CrossRef] [PubMed]

35. McLaren, M.E.; Nissim, N.R.; Woods, A.J. The effects of medication use in transcranial direct current stimulation: A brief review. *Brain Stimul.* **2017**, *11*, 52–58. [CrossRef] [PubMed]

36. Nitsche, M.A.; Kuo, M.-F.; Karrasch, R.; Wächter, B.; Liebetanz, D.; Paulus, W. Serotonin Affects Transcranial Direct Current-Induced Neuroplasticity in Humans. *Biol. Psychiatry* **2009**, *66*, 503–508. [CrossRef] [PubMed]

37. Kuo, H.-I.; Paulus, W.; Batsikadze, G.; Jamil, A.; Kuo, M.-F.; Nitsche, M.A. Chronic Enhancement of Serotonin Facilitates Excitatory Transcranial Direct Current Stimulation-Induced Neuroplasticity. *Neuropsychopharmacology* **2016**, *41*, 1223–1230. [CrossRef] [PubMed]

38. Brunoni, A.R.; Valiengo, L.; Baccaro, A.; Zanão, T.A.; de Oliveira, J.F.; Goulart, A.; Boggio, P.S.; Lotufo, P.A.; Benseñor, I.M.; Fregni, F. The sertraline vs. electrical current therapy for treating depression clinical study: Results from a factorial, randomized, controlled trial. *JAMA Psychiatry* **2013**, *70*, 383–391. [CrossRef] [PubMed]

39. D'Urso, G.; Dell'Osso, B.; Rossi, R.; Brunoni, A.R.; Bortolomasi, M.; Ferrucci, R.; Priori, A.; de Bartolomeis, A.; Altamura, A.C. Clinical predictors of acute response to transcranial direct current stimulation (tDCS) in major depression. *J. Affect. Disord.* **2017**, *219*, 25–30. [CrossRef] [PubMed]

40. Lewin, A.B.; de Nadai, A.S.; Park, J.; Goodman, W.K.; Murphy, T.K.; Storch, E.A. Refining clinical judgment of treatment outcome in obsessive–compulsive disorder. *Psychiatry Res.* **2011**, *185*, 394–401. [CrossRef] [PubMed]

**brain
sciences**

MDPI

Review

Depression, Olfaction, and Quality of Life: A Mutual Relationship

Marion Rochet [1], **Wissam El-Hage** [1,2], **Sami Richa** [3], **François Kazour** [1,4] and **Boriana Atanasova** [1,*]

[1]   UMR 1253, iBrain, Université de Tours, Inserm, 37200 Tours, France; marion.rochet@etu.univ-tours.fr (M.R.);
     wissam.elhage@univ-tours.fr (W.E.-H.); francoiskazour@hotmail.com (F.K.)
[2]   CHRU de Tours, Clinique Psychiatrique Universitaire, 37044 Tours, France
[3]   Department of Psychiatry, Faculty of Medicine, Saint-Joseph University, P.O. Box 17-5208,
     11-5076 Beirut, Lebanon; richasami@hotmail.com
[4]   Psychiatric Hospital of the Cross, 60096 Jal Eddib, Lebanon
[*]   Correspondence: atanasova@univ-tours.fr; Tel.: +33-2-47-36-73-05

Received: 13 April 2018; Accepted: 3 May 2018; Published: 4 May 2018

Abstract: Olfactory dysfunction has been well studied in depression. Common brain areas are involved in depression and in the olfactory process, suggesting that olfactory impairments may constitute potential markers of this disorder. Olfactory markers of depression can be either state (present only in symptomatic phases) or trait (persistent after symptomatic remission) markers. This study presents the etiology of depression, the anatomical links between olfaction and depression, and a literature review of different olfactory markers of depression. Several studies have also shown that olfactory impairment affects the quality of life and that olfactory disorders can affect daily life and may be lead to depression. Thus, this study discusses the links between olfactory processing, depression, and quality of life. Finally, olfaction is an innovative research field that may constitute a new therapeutic tool for the treatment of depression.

Keywords: depression; olfaction; markers; quality of life; therapeutic tool

1. Introduction

Depression and olfactory dysfunction induce long-term impairments that affect individuals and have a major impact on subjects' social skills, relationships, wellbeing, and quality of life [1,2]. Major depression is one of the two most debilitating diseases according to the World Health Organization (2010). Depression affects 8 to 12% of the world's population [3]. In France, 9 million people experience at least one depressive episode during their lifetime (INPES, National Institute for Prevention and Health Education). Olfactory dysfunctions are present in 22% of individuals aged between 25 and 75 years old [4]. In these situations, the impact of this olfactory impairment on patients' life is often neglected. However, olfaction provides people with valuable input from the chemical environment around them. Smell can also impact different psychological aspects of the subject's life by forming positive and negative emotional memories related to smell. It can also affect the social abilities and the interpersonal relationships of the individual. When the olfactory input is distorted, disability and decreased quality of life are reported [5].

In the last two decades, several studies have investigated olfactory dysfunction in depression. Several reasons are behind the study of olfaction in depression. First, the rapid expansion of structural and functional imagery techniques that uncover overlapping brain areas involved in olfactory processing and depression. These regions are mainly the orbitofrontal cortex, the anterior and posterior cingulate cortices, the insula, the amygdala, the hippocampus, and the thalamus [6–8]. Second, olfactory stimuli are encoded and may activate emotional memory. This could be explained by the close anatomical links between the olfactory system and the brain circuits involved in memory [9] and emotion [10]. These two cognitive functions are frequently altered in depression [11]. Third, olfactory dysfunction may induce symptoms of depression, due to the impact of odors on emotions, mood, or behaviors [12]. Indeed, a recent systematic review has shown a reciprocity and a close association between olfactory impairment and depression [13]. Thus, depressed patients have lower olfactory performance than healthy controls, and patients with olfactory dysfunction have symptoms of depression that worsen with the severity of olfactory impairment. Fourth, bilateral olfactory bulbectomy in rodents induces changes in behavior, as well as in the endocrine, immune, and neurotransmitter systems. These changes are similar to those seen in patients with major depression [14,15]. These alterations are reversed by antidepressants, suggesting that the dysfunctions seen in depression could be related to alterations in the olfaction system. However, these findings cannot be replicated in humans. Nevertheless, studies showed that olfactory bulb volume differs significantly between depressed and non-depressed individuals and seems to be a promising marker for depression [16]. Fifth, stress, a triggering factor of depression in vulnerable subjects, induces behavior similar to some symptoms of depression, as well as decreased cell proliferation or neurogenesis, both in the hippocampus and in the olfactive bulbs [17]. Finally, it has been suggested in human [18–21] and animal studies [22–24] that olfaction can be used as a therapeutic tool for depression.

The exact relationship between depression and olfactory dysfunction is still unclear, even if both disorders coexist and affect quality of life, social and professional integration, as well as family balance [25]. Several reviews have studied olfactory perception in depression [26,27], the relationship between olfactory loss and quality of life [28,29], and between depression and quality of life [1]. In the present review, we overview the links between depression, olfaction, and quality of life of patients in order to understand the relationship between olfactory function and depression, and to determine if olfactory loss affects patients' quality of life. In addition, we discuss the possible use of olfactory stimulation as a promising therapeutic tool to potentiate the antidepressant effect of medication.

2. Depression

Depression is a multifactorial disease involving multiple etiologies, with the contribution of biological, genetic, and environmental factors [30].

From a biological perspective, depression is associated to a monoaminergic deficiency in the brain [31]. Indeed, the role of monoaminergic neurotransmitters (noradrenaline, serotonin, and dopamine) has been demonstrated in the control of mood and cognitive functions. Neurotransmitter-based drug treatments accelerate clinical improvement, but resistance to antidepressants implies the involvement of other etiologies. One hypothesis is the inhibition of hippocampal neurogenesis, since long-term use of antidepressants leads to an increase of adult hippocampal neurogenesis [32]. A decrease in hippocampal volume has been demonstrated in depression [33] at the earliest stages of the disease and was correlated with patients' cognitive impairment [34]. Brain abnormalities have also been observed in the orbitofrontal cortex, anterior cingulate cortex [35], and the amygdala [36]. It has been shown that amygdala volume is reduced in untreated depressed patients, but increases with treatment [37].

As for the genetic etiologies of depression, studies show contradicting results. A meta-analysis suggested that depression is a familial disorder and with an estimated heritability of 31 to 42% [38]. For instance, a functional polymorphism of the serotonin transporter gene would change the impact of stressful life events on depression [39]. However, another meta-analysis was not able to show any association between the serotonin transporter genotype and depression [40]; and a genome-wide association study could not clearly identify the involvement of several genes in depression [41].

Concerning the psychological perspective, Beck [42] proposed a model based on the fact that negative life events may trigger depressive episodes. Therefore, certain life experiences can modify the psychological functioning of the individual and induce emotional instability. Kendler et al. [43] proposed the kindling-sensitization hypothesis where negative life events that were initially unable to trigger a depressive episode later gained a capacity to trigger a recurring episode.

Pathophysiology of depression is complex and remains partially elucidated. According to the latest edition of the Diagnostic and Statistical Manual (DSM-5, [44]), a major depressive episode is diagnosed, when (i) five (or more) of the following symptoms are present during the same two-week period: depressed mood most of the day, markedly diminished interest or pleasure in all, or almost all, activities most of the day, nearly every day, significant weight loss when not dieting or weight gain, insomnia or hypersomnia, psychomotor agitation or retardation, fatigue or loss of energy, feelings of worthlessness or excessive or inappropriate guilt, diminished ability to think or concentrate and recurrent thoughts of death; and (ii) a change from previous functioning with at least one of the two major symptoms of depression: depressed mood or loss of interest or pleasure. These symptoms are associated with significant clinical distress or impairment in social, occupational, or other important areas of functioning. It was shown that depression alters aspects of information processing, including perception and attention, memory processes (e.g., preferential recall of negative information rather than positive one), and interpretation of ambiguous information [11]. Recent literature explored the presence of depression-associated sensorial biases, particularly at the olfactory level by investigating the state (disappearance of olfactory alterations in clinically improved patients) and the trait (persistent olfactory alterations after clinical improvement) olfactory markers of major depression [27,45–48].

3. Olfaction

3.1. Functioning of Olfactory Perception

The olfactory information (i.e., the chemical molecule), will first settle at the level of the olfactory receptors, within the olfactory epithelium. There are two ways for volatile chemical molecules to reach the olfactory epithelium. The first is the orthonasal pathway (so-called "direct" pathway) and takes place during inspiration through the nose. The second is the retronasal pathway (so-called "indirect" pathway) and takes place during mastication through the mouth. The binding of the molecule on specific olfactory neuroreceptors, triggers an action potential on these cells [49]. The information then reaches the olfactory bulb, located at the base of the frontal lobe [2]. The olfactory bulb sends projections via the lateral olfactory tract to different brain areas, such as the olfactory tubercle, the anterior olfactory nucleus, the piriform cortex, the lateral entorhinal cortex, or the ventral tenia tectae [50], allowing olfactory information to reach different brain levels (e.g., the thalamus, hypothalamus, or hippocampus). Indeed, the olfactory system is connected to the limbic system through the amygdala, the piriform cortex, the anterior cingulate cortex, the insula, and the orbitofrontal cortex [7]. These connections may explain how odors an induce mood changes [51] and impact cognitions and behaviors [52].

3.2. Evaluation of Olfactory Functions

Various olfactory tests are available for the evaluation of olfactory function. Psychophysical measures have been used to establish a link between measurable parameters (i.e., product concentration, chemical composition) and the qualitative (i.e., odor's identification, odor's naming) and quantitative (i.e., perception of odor's intensity) characteristics of the evoked stimulus. It is, thus, possible to evaluate parameters, such as the odor's intensity, familiarity, or hedonic level by using scales of measurements. The psychophysical tests provide a quantitative measure of sensory function by using the verbal response of the subject as an indicator of the olfactory perception. Generally, the psychophysical tests include investigation of odor detection threshold, odor discrimination, and identification. Standardized tests have been developed to evaluate these three olfactory functions. They were validated with several hundred healthy participants (men and women) of different ages. Doty et al. [53] has developed the University of Pennsylvania Smell Identification Test (UPSIT), an olfactory evaluation tool in the form of a scratch 'n sniff test. This test can measure individuals' odor identification ability. The Sniffin' Stick Test created by Hummel et al. [54] evaluates odor detection threshold, odor discrimination, and identification ability. This test is formed of several odorous pens (sticks). For the olfactory threshold test, stick triplets with increasing odor concentrations are presented to the subject. In every triplet, only one stick contains the odorant (2-phenyl ethyl alcohol). The subject is asked to identify in every triplet the pen that contains the odorant, the other two being without odor. The discrimination test is done with the following presentation of 16 odorous pen triplets. In this test, the subject has to detect if the pen with the odorant is different from the other two. Finally, during the odor identification test, 12 or 16 odorous pens are presented to the subject. Using a multiple forced-choice task, odors are identified from a list of four descriptors for each odor. A final score evaluates the overall olfactory capacity of the subject and determines whether the individual is normosmic, hyposmic, or anosmic.

Occasionally, devices called olfactometers are used to carry out the psychophysical measures while providing a better control of the delivery conditions of the odorant. It allows the delivery of a precise concentration of the odorant at the entrance of the nostrils via a controlled airflow. It is also possible with this device to have either unilateral or bilateral nostril stimulations. This tool seems to be more appropriate in situations where the control of odors' concentration is important, like in olfactory threshold measures.

The psychophysical tests described above are easy to administer, to transport, are not expensive, and provide a rapid time of analysis (between 5 and 40 min depending on the test). They are the most used in the clinical field. However, the psychophysical evaluation of the olfactory functions demands an active participation of the subject. To avoid this inconvenience neurophysiological techniques are used to measure the human electrophysiologic response to an odorant stimulus. They include odor event-related potentials (OERPs) and a summated potential recorded from the surface of the olfactory epithelium (the electro-olfactogram, EOG). Many other techniques have also been developed, including electroencephalographic testing, measurement of psychogalvanic skin response to olfactory stimuli, measurement of respiratory, cardiovascular, papillary, and oculomotor reflexes [55]. More recently, structural and functional imaging technologies (positron emission tomography, single photon emission tomography, and high-resolution structural magnetic resonance imaging) were developed in order to study the central pattern associated to the olfactory perception.

In summary, a number of techniques and tools are available to explore the olfactory function, each having its own advantages and disadvantages. It is important to note that all these methods (psychophysical, neurophysiological, and neuroimaging techniques) are complementary because they do not measure the same olfactory parameters, function, and process. Some authors described and compared the existing methods and techniques (see [55,56]).

The methods described above can be used to study olfactory function and to detect the presence of olfactory disorders. Olfactory disorders are divided into quantitative and qualitative disorders. Quantitative olfactory disorders are hyposmia, hyperosmia, or anosmia [50]. They are defined, respectively, as a decrease, an increase, or a complete loss of olfactory perception. Qualitative olfactory disorders are parosmia (bad perception of smell) and fantosmia (an olfactory hallucination, or perception of odors that are not present within the olfactory field) [57]. Two thirds of cases of anosmia and hyposmia are due to either an upper respiratory tract infection, a brain trauma, or to nasal pathologies that damage the olfactory neuroepithelium [58]. Moreover, it has been demonstrated that olfactory perception is altered in patients with neurodegenerative diseases. Indeed, olfactory disorders are seen in 85 to 90% of patients with Alzheimer and Parkinson diseases. However, these patients are rarely aware of their olfactory impairment.

4. What Is the Link between Olfaction and Depression?

4.1. Anatomical Link

Several brain areas play a role in olfactory perception and are involved in the etiology of depression. First, the olfactory bulb transmits olfactory information to other brain areas, like the amygdala, the hippocampus, and the anterior angular cortex [50]. It has been shown that bilateral olfactory bulbectomy in rodents causes changes to the immune and endocrine systems similar to those seen in depression [15]. Indeed, bilateral destruction of olfactory bulbs leads to alteration in serotonin and dopamine concentration [14]. Bilateral olfactory bulbectomy can also induce behavioral changes, such as a reduction in sexual behavior [59], an odd food-motivated behavior [60], a decrease in anxiety-like behavior [61], and an increase in depression-like behavior [15,62]. In addition, a study found a reduced volume of the olfactory bulb in depressed patients [63]. A study using a rat model of depression, with unpredictable chronic mild stress (UCMS), has observed a reduced amount of olfactory receptor neurons in olfactory epithelium [64]. These results may explain some of the impairment in olfactory sensitivity observed in depressed patients.

A recent study showed that in a murine model that mimics the hyperactivation of the HPA axis (and thus causes a phenotype of depression), olfactory function, and also adult neurogenesis at the subventricular zone and the dentate gyrus, were affected [48]. In this study, mice receiving chronic administration of corticosterone had deficits in their olfactory acuity, fine odor discrimination, and olfactory memory. In addition, cell proliferation in the subventricular zone and the dentate gyrus (two niches of adult neurogenesis), and the survival of new neurons in the dentate gyrus and olfactory bulbs, were decreased by corticosterone administration. Antidepressant treatment (fluoxetine) allowed a return to normal olfactory function and adult neurogenesis [48]. Therefore, the link between olfaction and depression can also be explained by the decrease in neurogenesis.

Other areas, such as the amygdala or hippocampus, also have a role in olfaction and depression. Indeed, the hippocampus is involved in odor storage tasks [65] and in depressive symptoms, such as deficits in autobiographical memory [66]. In addition, studies have shown decreases in hippocampal volume associated to depression [33]. It has been shown that the amygdala of healthy individuals is activated during the evaluation of intensity, hedonic aspect and memory of odor-related emotions [67]. The amygdala would be hyper-activated in depression [68].

The orbitofrontal cortex is also implicated in the link between olfaction and depression. It is involved in attention, emotional, and cognitive processes of depression. On one hand, the ventromedial part is involved in rumination, anxiety, and sensitivity to pain, and is hyper-activated in depressed patients. On the other hand, the dorsal part is involved in psychomotor retardation, apathy, attention disorders, and working memory, and hypoactive in depressed patients [69]. The role of this cortex is crucial in olfaction, but is still controversial in depression [70,71]. Some authors consider that an unpleasant stimulus activates the left part of this cortex and a pleasant stimulus activates the right side [71], but other authors have shown that the activation of the orbitofrontal cortex is not a function of positive or negative valence of odorants [72]. The orbitofrontal cortex is involved in the identification of odors and in olfactory memorization [71,73].

The cingulate cortex is involved in both olfactory function and depression. In depression, the volume of its anterior part is diminished [35,74]. This would be partly responsible for the increased recurrence of depressive episodes [75]. As for its role in olfaction, the activation of this brain area depends on the hedonic valence of the odor [76].

The insula participates in the evaluation of emotional states and more particularly of the bodily sensations during an emotional experience [77]. A study showed that the insular cortex contributes to odor quality coding by representing the taste-like aspects of food odors [78]. The insula has higher levels of activity in resting states, increasing the inability of depressed patients to disengage from externally-cued events, and leading to pathological self-focused mental ruminative behaviors [79].

Finally, the habenula is affected by olfactory bulb input and is involved in the regulation of psychomotor and psychosocial behaviors [57]. Its metabolic activity is increased in animal models of depression [80]. The role of the habenula is the transfer of olfactory information to other brain areas [81]. It is activated in response to emotionally-negative stimuli [82]. According to Oral et al. [83] habenula plays a decisive role in the link between olfactory disorders and depression since a bilateral bulbectomy induces a structural degeneration by apoptosis of the habenula, leading to the appearance of the main symptoms of depression. A summary of the brains areas involved in olfaction and depression is presented in Table 1.

4.2. Olfaction: A Marker of Depression?

Over the years, several studies have focused on the impact of depression on olfactory functions like detection thresholds, identification, and discrimination capabilities, and the assessment of hedonicity or intensity. The strong link between depression and olfaction has allowed researchers to espouse the hypothesis that a reduced olfactory capacity may be a marker of depression [45,84]. Two types of olfactory markers have been proposed: (i) the state olfactory markers where olfactory impairments disappear after antidepressant treatment; and (ii) the trait olfactory marker, where olfactory impairments persist after clinical remission.

Most studies have shown that the detection threshold of depressed subjects was increased compared to controls [27,63,85,86]. However, some authors reported an unchanged olfactory threshold [87–89]. Few studies investigated the olfactory threshold in remitted patients after antidepressant treatment and showed conflicting results. Gross-Isseroff et al. [90] have demonstrated an increase of olfactory odor sensitivity in remitted patients suggesting that this could be due to antidepressant treatment. Another study observed a significant negative correlation between olfactory sensitivity and depressive symptoms [91]. Pause et al. [86] reported remission of odor threshold impairment in depressed patients after antidepressant treatment. All these observations suggest that the reduced olfactory sensitivity may be a marker of depression. However, further studies are needed to confirm whether this olfactory function is restored by antidepressant treatment.

Some studies have shown that depression is associated with a lower olfactory identification capacity [92–94]. Two studies reported that depressed subjects had a lower identification capacity of the components of a complex odorant environment (binary iso-intense mixtures), during the major depressive episode [46,95]. Naudin et al. [46] also showed that the olfactory alteration persists after clinical improvement reflecting olfactory trait markers of depression. The majority of studies have demonstrated that olfactory identification capacities are not altered in depression [46,85,87–89,91,96–100]. In summary, odor identification function didn't seem to be altered in depression when standardized olfactory tests were used. Moreover, it has been proposed that an odor identification parameter could be used to differentiate between depressed patients and Alzheimer's disease (AD) patients since comparative studies showed that this function is altered in AD, but not in depression (for review, see [8]).

Concerning odor hedonic perception, only two studies have shown unchanged scores between depressed patients and healthy controls [89] before or after antidepressant treatment [92]. For the majority of the investigations, this parameter is influenced by depressive state [46,47,85,86,95,101]. Atanasova et al. [95] showed that depressed patients would perceive unpleasant odors as more unpleasant (negative olfactory alliesthesia), while pleasant odorants would be perceived as less pleasant (olfactory anhedonia) in comparison to controls. Naudin et al. [46] reported that this hedonic olfactory bias concerns a highly emotional odorant and that it vanishes after antidepressant treatment. Therefore, it is considered as an olfactory state marker of depression. In contrast to these observations, other studies revealed that depressed patients over-evaluate the hedonic perception of odors [85,86]. Lombion-Pouthier et al. [85] suggested that this over-evaluation could be due to modifications in the orbitofrontal cortex observed in depression (this structure is also applied in hedonic perception of odor).

Several studies investigated changes in olfactory intensity rating in depression [85,86,92,95,102,103]. Only one showed that depressed patients perceived pleasant stimuli as less intense and unpleasant stimuli as more intense than controls [95]. However, the majority of the studies did not find any associations between that hedonic value of odors and changes in the perception of odor intensity.

As for the perception of familiarity, studies show contradictory results. Some authors could not find any change in familiarity ratings associated with depression [47,102,103], whereas others found lower familiarity ratings in depressed patients compared to controls [46]. Future studies are needed to clarify this aspect of olfaction. The edibility perception of odors has not been studied in depression. However, knowing that eating disorders are frequently observed in depression, this parameter should be a subject for future investigation.

Studies show that odor discrimination abilities are not different in depressed patients compared to controls [63,84,92,95]. These same studies found similar results with respect to odor intensity assessment capabilities. A summary of the studies exploring the deficits in the different olfactory functions in depressed patients and in clinically-improved patients are presented in Table 2.

Table 1. Brain areas involved in the processing of olfaction and depression.

Brain Areas	Olfaction		Depression	
	Observations	References	Observations	References
Olfactory bulb	• Transmission of olfactory information to other brain areas	Brand, 2001 [50]	• A bulbectomy causes changes (molecular and behavioral) similar to those of depression • Reduced volume observed in depression	Song and Leonard, 2005 [15] Negoias et al. 2010 [63]
Amygdala	• Activation during the evaluation of intensity, hedonic aspect and emotional memory related to odors	Pouliot and Jones-Gotman, 2008 [67]	• Hyperactivated in depression	Drevets, 2003 [68]
Hippocampus	• Role in odor storage tasks	Kesner et al. 2002 [66]	• Deficit in autobiographical memory • Reduced volume observed in depression	Lemogne et al. 2006 [46] Campbell et al. 2004 [33]
Orbitofrontal cortex	• Controversial data: left part activated by unpleasant stimulus and right part activated by pleasant stimulus *vs.* activation not dependent of hedonic valence • Identification of odors and memorization	Grabenhorst et al. 2011 [72] Zald and Pardo, 1997 [71] Zald et al. 2002 [73]	• Ventromedian part (hyperactivated) involved in rumination anxiety and sensitivity to pain and dorsal part (hypoactivated) psychomotor retardation, apathy, attention disorders, and working memory	Rogers et al. 2004 [69]
Cingulate cortex	• Activation dependent of hedonic valence of the odor	Fulbright et al. 1998 [76]	• Reduced volume of anterior part: partly responsible for the increased recurrence of depressive episodes	Phillips et al. 2003 [74] van Tol et al. 2010 [35] Bhagwagar et al. 2008 [75]
Insula	• Odor quality coding	Veldhuizen et al. 2010 [78]	• Increased of activity during states of rest can be involved in inability to disengage from external events and lead to rumination	Silz and Hayley, 2012 [79]
Habenula	• Transfer olfactory information to other brain areas • Activated in response to emotionally-negative stimuli	Da Costa et al. 1997 [81] Hikosaka et al. 2008 [82]	• Metabolic activity increased	Shumake et al. 2003 [80]

Table 2. Summary of publications on olfactory function in depression.

Study	Number of Patients Female (f)/Male (m)	Mean Age ± Standard Deviation	Measures and Tools	Conclusions
Swiecicki et al. (2009) [89]	D: 46 (f) C: 30 (f)	D: 38.2 ± 1.6 C: 35.4 ± 2.1	One testing session Sniffin' Stick Test: • Odor threshold • Odor identification • Odor characterization: pleasant/unpleasant/neutral	• Intact odor threshold • Intact identification abilities • Intact odor hedonicity
Scinska et al. (2008) [88]	D: 25 (17/7) C: 60 (38/22)	D: 67.2 ± 1.2 C: 66.7 ± 0.9	One testing session Sniffin' Stick Test: • Odor threshold (n-butanol) • Odor identification	• Intact odor threshold • Intact identification abilities
Postolache et al. (1999) [87]	D: 24 C: 24	D: 42.8 ± 9.7 C: 42.1 ± 11.8	One testing session • Odor threshold: single stair-case procedure (PEA) • Identification: L-UPSIT and R-UPSIT	• Intact odor threshold • Intact identification abilities
Negoias et al. (2010) [63]	D: 21 (17/4) C: 21 (15/6)	D: 36.9 ± 10.1 C: 39.6 ± 11.4	One testing session Sniffin' Stick test: • Odor threshold (PEA) • Odor discrimination • Odor identification	• Modified odor threshold • Intact odor discrimination ability • Intact odor identification ability
Lombion-Pouthier et al. (2006) [85]	D: 49 (35/14) C: 58 (36/22)	D: 43.4 ± 17.5 C: 38.4 ± 13.9	One testing session Olfactory test (Ezus) • Odor sensibility • Detection/evaluation abilities (16 odors) • Odor Identification • Intensity rating • Hedonic rating	• Modified odor threshold • Intact detection/evaluation abilities • Intact identification abilities • Intact odor intensity • Modified odor hedonicity
Pause et al. (2001) [86]	D: 24 (15/9) C: 25 (15/9)	D: 48.4 ± 13.2 C: 44.2 ± 12.6	Two testing sessions (S1 and S2) • Odor threshold (PEA and eugenol) • Subjective odors rating (intensity and hedonicity, 10 odors)	• Modified odor threshold (S1) • Modified odor hedonicity (S1) and intact odor intensity (S1/S2)

Table 2. *Cont.*

Study	Number of Patients Female (f)/Male (m)	Mean Age ± Standard Deviation	Measures and Tools	Conclusions
Pentzek et al. (2007) [99]	C: 20 (15/5) C: 30 (24/6)	D: 73.4 ± 5.6 C: 77.1 ± 6.8	One testing session Sniffin' Stick Test • Odor identification	• Intact identification abilities
McCaffrey et al. (2000) [98]	D: 20 (11/9) A: 20 (13/7)	D: 67.55 ± 7.29 A: 74.15 ± 7.86	One testing session • Pocket Smell Test	• Intact identification abilities
Solomon et al. (1998) [100]	D: 20 (13/7) A: 20 (12/8)	D: 69.4 ± 7.7 A: 74.5 ± 7.7	One testing session • Pocket Smell test	• Intact identification abilities
Kopala et al. (1994) [97]	D: 21 (13/8) C: 77 (47/30)	D: 37.0 ± 9.6 C: 32.5 ± 11.1	One testing session UPSIT • Odor identification	• Intact identification abilities
Amsterdam et al. (1987) [96]	D: 51 (34/17) C: 51 (34/17)	D: 43 ± 13 (f); 49 ± 14 (m) C: age-matched	One testing session UPSIT • Odor identification	• Intact identification abilities
Naudin et al. (2012) [46]	D: 20 (12/6) C: 54 (36/18)	D: 50.1 ± 13.3 C: 49.5 ± 12.5	Two testing sessions (during acute phase of depression and after 6 weeks of antidepressant treatment) • Eight odors (four pleasant, two unpleasant, two neutral) • Hedonic rating • Familiarity rating • Identification of single odors • Odor intensity discrimination • Odor identification abilities in binary mixture (one pleasant/one unpleasant)	• Modified odor hedonicity (S1) and improvement in (S2) • Modified odor familiarity (S1, S2 for vanillin) • Intact identification abilities of single odors (S1, S2) • Modified odor intensity discrimination (S1, S2) • Modified identification abilities in binary mixture (S1, S2)
Zucco and Bollini (2011) [94]	Mild D: 12 (6/6) Severe D: 12 (6/6) C: 12 (6/6)	Mild D: 41.3 ± 6.4 Severe D: 41.9 ± 6.2 C: 39.8 ± 7.1	One testing session • Odor identification • Odor recognition memory (10 targets odors/30 distractors odors)	• Modified odor identification abilities • Modified odor recognition memory

Table 2. *Cont.*

Study	Number of Patients Female (f)/Male (m)	Mean Age ± Standard Deviation	Measures and Tools	Conclusions
Atanasova et al. (2010) [95]	D: 30 (12/18) C: 30 (12/18)	D: 34.6 ± 11.1 C: 33.4 ± 9.9	One testing session • Two odors: vanillin/butyric acid • Intensity rating • Odor intensity discrimination • Odor identification in binary mixture • Hedonics rating (single/binary mixture) • Odor discrimination	• Modified odor intensity • Modified odor intensity discrimination • Modified identification abilities in binary mixture • Modified odor hedonicity • Intact odor discrimination ability for unpleasant odor and modified for pleasant odor
Clepce et al. (2010) [92]	D: 37 (21/16) C: 37 (21/16)	D: 48.3 ± 11.9 (m); 47.5 ± 11.3 (f) C: age-matched	Two testing sessions (during acute phase of depression and six months after the 1st phase) Sniffin' Stick Test: • Odor identification • Hedonics and intensity rating	• Modified odor identification abilities (S1) • Intact odor hedonicity and intensity (S1, S2)
Pollatos et al. (2007) [91]	D: 48 (34/14)	28.2 ± 5.8	One testing session Sniffin' Stick test: • Odor threshold • Odor discrimination	• Modified odor threshold • Intact odor quality discrimination
Croy et al. (2014) [27]	D: 27 (f) C: 28 (f)	D: 38.5 ± 10.6 C: 353 ± 10.3	Two testing sessions (before and after therapy) Sniffin' Stick test: • Odor threshold • Odor discrimination • Odor identification	• Intact odor threshold (S1, S2) • Modified odor discrimination (S1, S2) • Intact odor identification (S1, S2)
Atanasova et al. (2012) [101]	D: 30 (12/18) C: 30 (12/18)	D: 36.6 ± 11.1 C: 33.4 ± 9.9	One testing session • Odor: vanillin/butyric acid • Hedonicity	• Modified odor hedonicity

Table 2. *Cont.*

Study	Number of Patients Female (f)/Male (m)	Mean Age ± Standard Deviation	Measures and Tools	Conclusions
Naudin et al. (2014) [47]	D: 22 (16/6) C: 41 (24/17)	D: 33.2 ± 11.2 C: 34 ± 11	Two testing sessions (one during acute phase of depression and 100 days ± 64 after the 1st phase) • Eight odors • Hedonicity, Familiarity and emotion' intensity rating	• Modified odor hedonicity (S1; restore in S2), intact odor familiarity (S1, S2) and modified intensity of emotion (S1)
Naudin et al. (2014) [102]	D: 20 (15/5) A: 20 (14/6) C: 24 (17/9)	D: 64.9 ± 11.2 A: 73.0 ± 11.2 C: 67.0 ± 12.7	One testing session (in elderly) • Eight odors: four familiar and four unfamiliar • Odor recognition memory • Pleasantness, intensity, familiarity	• Modified odor recognition memory for D and A • Intact odors pleasantness, intensity and familiarity perception for D and A
Gross-Isseroff et al. (1994) [90]	D: 9 (8/1) C: 9 (8/1)	D: 49 ± 4.6 C: 49.1 ± 4.8	Three testing sessions (1st: acute phase of depression; 2nd: three weeks after the 1st and 3rd: sox weeks after the 1st) • Odors threshold (androsteron and isoamylacetate)	• Intact odor threshold (1st and 2nd sessions) and modified odor threshold (3rd session)
Serby et al. (1990) [93]	D: 9 (m) C: 9 (m)	D: 50–59 years old C: age-matched	One testing phase • UPSIT (identification abilities) • Odor threshold (geraniol)	• Modified odor identification abilities • Intact odor threshold
Thomas et al. (2002) [103]	D: 16 C: 16	*i.i.*	• Odor threshold • Intensity and familiarity ratings	• Intact odor threshold • Intact intensity and familiarity ratings abilities
Postolache et al. (2002) [104]	SAD: 14 (7/7) C: 16 (9/7)	SAD: 42.3 ± 11.5 C: 39.0 ± 10.0	One testing session • Odor threshold	• Intact odor threshold
Oren et al. (1995) [105]	SAD: 21 (16/9) C: 21 (16/9)	SAD: 38 ± 9 C: 38 ± 9	One testing session • Odor identification	• Intact odor identification abilities
Kamath et al. (2017) [106]	BPI: 43 (28/15) BPII: 48 (29/19) D: 134 (97/34) Anxiety: 48 (36/12) C: 72 (30/42)	BPI: 43.66 ± 15.6 BPII: 44.85 ± 15.6 D: 50.31 ± 14.87 Anxiety: 48.78 ± 17.12 C: 50.50 ± 18.09	One testing session • UPSIT (identification abilities) • Sniffin' Stick Test: Discrimination abilities and odor threshold	• Modified odor identification abilities (BPI, D) • Intact discrimination abilities and odor threshold

A: Alzheimer patients; C: control group; D: depressed patients; SAD: seasonal affective disorder; *i.i.*: incomplete information; BP: bipolar disorder; BPI: bipolar depression; PEA: phenyl ethyl alcohol; UPSIT: University of Pennsylvania Smell Identification Test, S1: testing session during acute phase of depression, S2: testing session during remission and/or after therapy.

Olfactory impairments were also described in other psychiatric, disorders like schizophrenia [107], bipolar disorder [108], or posttraumatic stress disorder [109]. However, there are very few studies comparing olfactory dysfunction in depression and other affective disorders. When assessing olfactory thresholds and identification abilities, Swiecicki et al. [89] could not find any difference between patients with unipolar and bipolar depression. However, a recent study [106], showed that impairment in odor identification may be seen in mood disorders (major depressive and bipolar I disorder), with a more pronounced impairment associated to psychotic depression features. On the other hand, global olfactory dysfunction observed in schizophrenia may not be a feature of other neuropsychiatric conditions [106]. Olfactory markers of differentiation were also described in depression and Alzheimer's disease [8]. Future investigations are needed in order to investigate potential markers of differentiation between other neuropsychiatric disorders.

5. Quality of Life: Definition and Evaluation

The World Health Organization, defined the quality of life as "the perception of an individual of his place in existence, in the context of the culture and the system of values in which he lives, in relation to its objectives, expectations, norms and concerns. It is a very broad concept influenced in a complex way by the physical health of the subject, his psychological state, his level of independence, his social relations as well as his relation to the essential elements of its environment". Questionnaires have been created to assess the quality of life, like the Questionnaire of Olfactory Disorders [25]. This questionnaire was developed to assess the impact of olfactory disorders in daily life and is composed of 52 statements divided into three areas: (1) "negative statement", which provides information on the degree of suffering of the respondent; (2) "positive statement", that gives information on the way respondents cope with their olfactory disorders; and (3) "socially desired". This third area allows the evaluator to assess the credibility of participants' responses. Indeed, it helps knowing if the participant is trying to give a certain impression or socially desirable answers.

Other questionnaires used to assess quality of life include the "Short Form 36 Health Survey", "The General Well Being Schedule" [110], "The 90-item symptom checklist" [111], and "The Nottingham Health Profile" [112]. In these different questionnaires, several areas, like general health, vitality, depression, anxiety, sleep, and pain are evaluated. However, sensory functions, like taste and olfaction, are not evaluated. A study by MacDowell et al. [113] compared the reliability, validity, and duration needed for several of these instruments. They concluded that most of these tools perform adequately for survey research purposes.

This is a non-exhaustive list of standardized questionnaires used to assess quality of life, but researchers may prefer to use other non-standardized questionnaires targeting specific fields of research (olfactory disorders, etc.). For example, some studies used a self-report where participants had to write their feeling about their loss of olfaction [5,114].

6. Impact of Olfactory Disorders on Quality of Life

It is interesting to focus on the impact of olfactory loss (partial or total, reversible or not) on the diet in general. Individuals with olfactory disorders report having a more complicated alimentation because of the important link between smell and taste. Patients with olfactory disorders (69%, n = 239) had lower ratings of food since the beginning of their disorder [115]. This lower rating leads to a decrease in appetite in 56% (n = 278) [114], 32% (n = 72) [116] and 27% of subjects (n = 50) [117]. All the subjects included in these studies had olfactory disorders and the results were obtained through self-reporting questionnaires and the Multi-Clinic Smell and Taste Questionnaire [118].

Individuals sometimes develop coping strategies: they can increase the taste by adding salt, sugar, or spices [5,28]. Moreover, 3 to 20% of individuals with olfactory disorders report eating more than before, while from 20 to 36% report eating less [115]. The presence of olfactory dysfunction causes difficulty in maintaining a healthy balanced diet. Individuals with olfactory disorders also have difficulties cooking food because they have difficulties to detect burning food. They may burn or stale food because of their olfactory impairment [5,114,119,120]. They also have difficulties to detect gas leaks or smoke odors [120].

In addition, patients are no longer able to detect their own body odor (sweating, bad breath) [114] and report problems of hygiene. This can lead in some cases to excessive showers in order to have a higher self-confidence. For some individuals, this is the most negative effect of olfactory loss [116,117]. Between 1/4 and 1/3 of patients with olfactory disorders report having social difficulties related to hygiene problems [117,121]. Indeed, an individual with olfactory disorders is socially vulnerable if he is not able to smell his body odor or odors of others. This places him in an uncomfortable situation and increases the risk of social isolation.

Olfactory disturbance interferes with professional life in 3 to 8% of cases [116,117]. Temmel et al. [114] reported that 8% of participants with olfactory disorders had problems in their professional lives. The impact of an olfactory disorder on professional life depends on the type profession: people who work in oenology, in gastronomy, in the perfume industry, or even nurses or firefighters, can have major impairments in their professional lives [28]. A loss of smell can also be problematic in this kind of situation and add major concerns for the future professional. All of this can lead to anxiety, mood disturbances, and depression [58]. Olfactory disorders that cause food disturbance (whether cooking or eating) can also constitute factors that affect mood and lead to anhedonia [28].

Several studies have shown the impact of olfaction on social and family relations [122,123] such as the mother-child bond [124] and men-women relationships (i.e., in reproduction, avoidance of consanguinity, selection of partner) [125,126]. For example, olfaction is involved in the detection of fear signals. In a study by Ackerl et al. [122], women watched a terrifying movie while wearing axillary pads. A neutral film was presented the next day as control. Other women then had to smell and categorize the axillary pads obtained after the presentation of the terrifying film and those obtained after the presentation of the neutral film. As a result, women were able to distinguish between the pads of fear and those of non-fear. These results assume the existence of an odor of fear, present in perspiration and detectable by other individuals. Such behaviors may be also affected by olfactory disorders, but further studies still need to be done in this area.

7. Olfaction: A New Therapeutic Tool?

7.1. Animal Model and Odorants

Studies in animals have shown effects of odors on the emotional state. Komiya et al. [24] have demonstrated the anti-stress effects of lemon oil vapor on mice during behavioral tests (elevated plus maze and forced swim tests). The mechanism hypothesis is that lemon essential oil vapor affects the response to dopaminergic activity by modulating serotoninergic activity and/or GABA-benzodiazepines receptor complex.

In another study, Xu et al. [23] showed the effect of vanillin on the depressive-like behavior of non-bulbectomized and bulbectomized groups of rats. This was a comparative study of the effects of vanillin on depressive-like behavior rats induces by two ways: chronic unpredictable mild stress (CUMS) and olfactory bulbectomy. Results showed a significant decrease of depressive-like behavior for the CUMS group exposed to vanillin and the CUMS group exposed to fluoxetine. No changes were observed for the bulbectomy group exposed to vanillin. These results suggest that vanillin may have an effect on the symptoms of depression if the olfactory pathway is intact [23]. A study on Mongolian gerbil (which has more neuroendocrine similarity with humans than mice and rats) highlighted the anxiolytic effect of lavender odor. The authors showed that after chronic exposure to this odor, the anxiolytic effects were almost similar to those obtained after exposure to diazepam [22], but the mechanisms were not completely elucidated. Another study on rats demonstrated the antidepressant effect of lemon odor during behavioral tests (forced swim test and open field test) [127]. Lemon odor significantly reduced the depressive-like behavior. Further studies on the effects of odors on affective states in animal models are needed in order to understand the underlying mechanisms of these effects.

However, it is important to note that the cited studies have used bulbectomy to develop an animal model of depression. Their validity was, however criticized (for review, see [128]). Therefore, it is important to be careful before applying the findings obtained in animal models onto human subjects.

7.2. Use of Odors in Humans

Some studies have highlighted the sedative [129] and the anxiolytic effects [130,131] of odors in humans. Lehrner et al. [129] showed that participants exposed to the ambient odors of orange essential oils had lower levels of anxiety compared to control participants. The same authors extended their observations by using lavender odor [131]. The objective of this last study was to compare the effect of odor to the effect of music in the waiting room of a dental office. The results showed that the lavender odor had a more pronounced anxiolytic effect compared to music and control conditions. Indeed, studies showed that lavender acts post-synaptically and modulates the activity of cyclic adenosine monophosphate (cAMP). A reduction in cyclic adenosine monophosphate activity is associated with sedation [132].

Studies in humans have highlighted the interest of using odors as therapeutic tool. In a study by Hummel et al. [21], participants with olfactory loss (various origins: post-infectious, post-traumatic or idiopathic) followed a 12-week olfactory training program. At the end of this training, measurements of olfactory function were performed and it appeared that repeated daily olfactory stimulation improved the olfactory function of participants.

Haehner et al. [19] has proposed olfactory training in patients with Parkinson's disease. They used the same protocol as Hummel et al. [21]. The results of the study showed that olfactory training produced an increased olfactory sensitivity for the four odors used during the training, but also an overall increase of the olfactory function, whereas no anti-Parkinsonian treatment allowed this type of result.

Finally, a recent study highlighted the effects of olfactory training on symptoms of depression [20]. In this study, the subjects followed an olfactory training over a period of five months or had to perform Sudoku daily for the control group. The odors used were the same as in Hummel's study presented previously [21]. The results showed a significant decrease in the depression score (obtained with the Beck Depression Inventory) for the group who followed the five-month olfactory training compared to the group who performed Sudoku daily.

The mechanisms of action of odors' effects on depression are not known yet. However, several hypotheses were developed: the volatile odorant compound could act as a pharmacological agent and enter the bloodstream. In such a situation, its effects will be dependent on the concentration of the compound [133]. Studies are still needed to understand the mechanisms of action of odors on behavior in animals or humans.

8. Conclusions

We have shown here the links between the olfactory system, depression, and quality of life. Different brain areas are involved in both depression and olfaction, and patients with depression regularly suffer from an impaired sense of smell. Further studies are needed to confirm that only olfactory threshold and hedonic perception are altered in depression while the odor identification capabilities are preserved. In the future, it is important to study the olfactory perception of depressed patients in a more natural environment reflecting everyday life and using more complex sensory (olfactory and gustatory) stimuli. These investigations could explain the role of olfactory impairment in the eating disorders frequently observed in depression.

In this review, we have also demonstrated that the presence of olfactory disorders can lead to a decrease in the quality of life of patients in several areas: food, social life, and work. Previous studies have demonstrated that other perceptual deficits affecting gustation [134], vision [135], or audition [136] may lead to depression by decreasing quality of life. There are no comparative studies investigating the importance of different perceptual sensory deficits on depressed patients' quality of life. However, it is known that all senses participate in some sensory experiences like food intake. Such sensory experience will involve vision, audition, and kinesthesis, as well as the chemosensory modalities of olfaction, gustation, and chemesthesis that underlies flavor perception. However, sensory deficits are only partially assessed in some of the questionnaires evaluating the quality of life. Future improvement of the existing questionnaires are needed to create the tools suitable to the real problems of the clinical populations with olfactory deficits. Moreover, current clinical practice does not take into account olfactory impairments in depressed patients; olfactory deficits are not described in depression clinical criteria defined in DSM-5 [44]. The present review confirms the presence of sensory impairments and, specifically, the olfactory ones in the clinical spectrum of depression patients, and suggests that simple and inexpensive tools could be used to improve olfactory deficits. Furthermore, the current literature provides emerging evidence that olfactory stimulation (olfactory training) may be a promising tool for future therapeutic prospects. As suggested by Hummel et al. [21], it would be interesting to study the effects of this stimulation over time in order to know if the changes observed during olfactory training persist in time or not. Finally, studying the precise mechanisms of action of olfactory training is a must, in order to develop better olfactory training methods and to deepen knowledge about the olfactory system, depression, and how they affect quality of life.

Author Contributions: M.R., B.A., W.E.-H., F.K. and S.R. wrote the paper.

Conflicts of Interest: The authors declare no conflict of interest.

References

1. Sivertsen, H.; Bjørkløf, G.H.; Engedal, K.; Selbæk, G.; Helvik, A.-S. Depression and Quality of Life in Older Persons: A Review. *Dement. Geriatr. Cogn. Disord.* **2015**, *40*, 311–339. [CrossRef] [PubMed]
2. Doty, R. The Olfactory System and Its Disorders. *Semin. Neurol.* **2009**, *29*, 74–81. [CrossRef] [PubMed]
3. Andrade, L.; Caraveo-Anduaga, J.J.; Berglund, P.; Bijl, R.V.; De Graaf, R.; Vollebergh, W.; Dragomirecka, E.; Kohn, R.; Keller, M.; Kessler, R.C.; et al. The epidemiology of major depressive episodes: Results from the International Consortium of Psychiatric Epidemiology (ICPE) Surveys. *Int. J. Methods Psychiatr. Res.* **2003**, *12*, 3–21. [CrossRef] [PubMed]
4. Vennemann, M.M.; Hummel, T.; Berger, K. The association between smoking and smell and taste impairment in the general population. *J. Neurol.* **2008**, *255*, 1121–1126. [CrossRef] [PubMed]
5. Miwa, T.; Furukawa, M.; Tsukatani, T.; Costanzo, R.M.; DiNardo, L.J.; Reiter, E.R. Impact of olfactory impairment on quality of life and disability. *Arch. Otolaryngol. Head Neck Surg.* **2001**, *127*, 497–503. [CrossRef] [PubMed]
6. Höflich, A.; Baldinger, P.; Savli, M.; Lanzenberger, R.; Kasper, S. Imaging treatment effects in depression. *Rev. Neurosci.* **2012**, *23*, 227–252. [CrossRef] [PubMed]
7. Soudry, Y.; Lemogne, C.; Malinvaud, D.; Consoli, S.-M.; Bonfils, P. Olfactory system and emotion: Common substrates. *Eur. Ann. Otorhinolaryngol. Head Neck Dis.* **2011**, *128*, 18–23. [CrossRef] [PubMed]
8. Naudin, M.; Atanasova, B. Olfactory markers of depression and Alzheimer's disease. *Neurosci. Biobehav. Rev.* **2014**, *45*, 262–270. [CrossRef] [PubMed]
9. Savic, I.; Gulyas, B.; Larsson, M.; Roland, P. Olfactory functions are mediated by parallel and hierarchical processing. *Neuron* **2000**, *26*, 735–745. [CrossRef]
10. Anderson, A.K.; Christoff, K.; Stappen, I.; Panitz, D.; Ghahremani, D.G.; Glover, G.; Gabrieli, J.D.E.; Sobel, N. Dissociated neural representations of intensity and valence in human olfaction. *Nat. Neurosci.* **2003**, *6*, 196–202. [CrossRef] [PubMed]
11. Kircanski, K.; Joormann, J.; Gotlib, I.H. Cognitive aspects of depression: Cognitive aspects of depression. *Wiley Interdiscip. Rev. Cogn. Sci.* **2012**, *3*, 301–313. [CrossRef] [PubMed]
12. Seo, H.-S.; Roidl, E.; Müller, F.; Negoias, S. Odors enhance visual attention to congruent objects. *Appetite* **2010**, *54*, 544–549. [CrossRef] [PubMed]

13. Kohli, P.; Soler, Z.M.; Nguyen, S.A.; Muus, J.S.; Schlosser, R.J. The Association between Olfaction and Depression: A Systematic Review. *Chem. Senses* **2016**, *41*, 479–486. [CrossRef] [PubMed]

14. Masini, C.V.; Holmes, P.V.; Freeman, K.G.; Maki, A.C.; Edwards, G.L. Dopamine overflow is increased in olfactory bulbectomized rats: An in vivo microdialysis study. *Physiol. Behav.* **2004**, *81*, 111–119. [CrossRef] [PubMed]

15. Song, C.; Leonard, B.E. The olfactory bulbectomised rat as a model of depression. *Neurosci. Biobehav. Rev.* **2005**, *29*, 627–647. [CrossRef] [PubMed]

16. Rottstaedt, F.; Weidner, K.; Strauß, T.; Schellong, J.; Kitzler, H.; Wolff-Stephan, S.; Hummel, T.; Croy, I. Size matters—The olfactory bulb as a marker for depression. *J. Affect. Disord.* **2018**, *229*, 193–198. [CrossRef] [PubMed]

17. Mineur, Y.S.; Belzung, C.; Crusio, W.E. Functional implications of decreases in neurogenesis following chronic mild stress in mice. *Neuroscience* **2007**, *150*, 251–259. [CrossRef] [PubMed]

18. Croy, I.; Hummel, T. Olfaction as a marker for depression. *J. Neurol.* **2017**, *264*, 631–638. [CrossRef] [PubMed]

19. Haehner, A.; Tosch, C.; Wolz, M.; Klingelhoefer, L.; Fauser, M.; Storch, A.; Reichmann, H.; Hummel, T. Olfactory Training in Patients with Parkinson's Disease. *PLoS ONE* **2013**, *8*, e61680. [CrossRef] [PubMed]

20. Wegener, B.-A.; Croy, I.; Hähner, A.; Hummel, T. Olfactory training with older people: Olfactory training. *Int. J. Geriatr. Psychiatry* **2018**, *33*, 212–220. [CrossRef]

21. Hummel, T.; Rissom, K.; Reden, J.; Hähner, A.; Weidenbecher, M.; Hüttenbrink, K.-B. Effects of olfactory training in patients with olfactory loss. *Laryngoscope* **2009**, *119*, 496–499. [CrossRef] [PubMed]

22. Bradley, B.F.; Starkey, N.J.; Brown, S.L.; Lea, R.W. Anxiolytic effects of *Lavandula angustifolia* odour on the Mongolian gerbil elevated plus maze. *J. Ethnopharmacol.* **2007**, *111*, 517–525. [CrossRef] [PubMed]

23. Xu, J.; Xu, H.; Liu, Y.; He, H.; Li, G. Vanillin-induced amelioration of depression-like behaviors in rats by modulating monoamine neurotransmitters in the brain. *Psychiatry Res.* **2015**, *225*, 509–514. [CrossRef] [PubMed]

24. Komiya, M.; Takeuchi, T.; Harada, E. Lemon oil vapor causes an anti-stress effect via modulating the 5-HT and DA activities in mice. *Behav. Brain Res.* **2006**, *172*, 240–249. [CrossRef] [PubMed]

25. Frasnelli, J.; Hummel, T. Olfactory dysfunction and daily life. *Eur. Arch. Otorhinolaryngol.* **2005**, *262*, 231–235. [CrossRef] [PubMed]

26. Taalman, H.; Wallace, C.; Milev, R. Olfactory Functioning and Depression: A Systematic Review. *Front. Psychiatry* **2017**, *8*. [CrossRef] [PubMed]

27. Croy, I.; Symmank, A.; Schellong, J.; Hummel, C.; Gerber, J.; Joraschky, P.; Hummel, T. Olfaction as a marker for depression in humans. *J. Affect. Disord.* **2014**, *160*, 80–86. [CrossRef] [PubMed]

28. Croy, I.; Nordin, S.; Hummel, T. Olfactory Disorders and Quality of Life—An Updated Review. *Chem. Senses* **2014**, *39*, 185–194. [CrossRef] [PubMed]

29. Hummel, T.; Nordin, S. Olfactory disorders and their consequences for quality of life. *Acta Otolaryngol. (Stockh.)* **2005**, *125*, 116–121. [CrossRef]

30. El Hage, W.; Powell, J.F.; Surguladze, S.A. Vulnerability to depression: What is the role of stress genes in gene x environment interaction? *Psychol. Med.* **2009**, *39*, 1407–1411. [CrossRef] [PubMed]

31. Willner, P.; Scheel-Krüger, J.; Belzung, C. The neurobiology of depression and antidepressant action. *Neurosci. Biobehav. Rev.* **2013**, *37*, 2331–2371. [CrossRef] [PubMed]

32. Malberg, J.E.; Eisch, A.J.; Nestler, E.J.; Duman, R.S. Chronic antidepressant treatment increases neurogenesis in adult rat hippocampus. *J. Neurosci. Off. J. Soc. Neurosci.* **2000**, *20*, 9104–9110. [CrossRef]

33. Campbell, S.; Marriott, M.; Nahmias, C.; MacQueen, G.M. Lower hippocampal volume in patients suffering from depression: A meta-analysis. *Am. J. Psychiatry* **2004**, *161*, 598–607. [CrossRef] [PubMed]

34. Hickie, I.; Naismith, S.; Ward, P.B.; Turner, K.; Scott, E.; Mitchell, P.; Wilhelm, K.; Parker, G. Reduced hippocampal volumes and memory loss in patients with early- and late-onset depression. *Br. J. Psychiatry J. Ment. Sci.* **2005**, *186*, 197–202. [CrossRef] [PubMed]

35. Van Tol, M.-J.; van der Wee, N.J.A.; van den Heuvel, O.A.; Nielen, M.M.A.; Demenescu, L.R.; Aleman, A.; Renken, R.; van Buchem, M.A.; Zitman, F.G.; Veltman, D.J. Regional brain volume in depression and anxiety disorders. *Arch. Gen. Psychiatry* **2010**, *67*, 1002–1011. [CrossRef] [PubMed]

36. Kronenberg, G.; Tebartz van Elst, L.; Regen, F.; Deuschle, M.; Heuser, I.; Colla, M. Reduced amygdala volume in newly admitted psychiatric in-patients with unipolar major depression. *J. Psychiatr. Res.* **2009**, *43*, 1112–1117. [CrossRef] [PubMed]

37. Hamilton, J.P.; Siemer, M.; Gotlib, I.H. Amygdala volume in major depressive disorder: A meta-analysis of magnetic resonance imaging studies. *Mol. Psychiatry* **2008**, *13*, 993–1000. [CrossRef] [PubMed]
38. Sullivan, P.F.; Neale, M.C.; Kendler, K.S. Genetic epidemiology of major depression: Review and meta-analysis. *Am. J. Psychiatry* **2000**, *157*, 1552–1562. [CrossRef] [PubMed]
39. Caspi, A.; Sugden, K.; Moffitt, T.E.; Taylor, A.; Craig, I.W.; Harrington, H.; McClay, J.; Mill, J.; Martin, J.; Braithwaite, A.; et al. Influence of life stress on depression: Moderation by a polymorphism in the 5-HTT gene. *Science* **2003**, *301*, 386–389. [CrossRef] [PubMed]
40. Risch, N.; Herrell, R.; Lehner, T.; Liang, K.-Y.; Eaves, L.; Hoh, J.; Griem, A.; Kovacs, M.; Ott, J.; Merikangas, K.R. Interaction between the serotonin transporter gene (5-HTTLPR), stressful life events, and risk of depression: A meta-analysis. *JAMA* **2009**, *301*, 2462–2471. [CrossRef] [PubMed]
41. Hek, K.; Demirkan, A.; Lahti, J.; Terracciano, A.; Teumer, A.; Cornelis, M.C.; Amin, N.; Bakshis, E.; Baumert, J.; Ding, J.; et al. A genome-wide association study of depressive symptoms. *Biol. Psychiatry* **2013**, *73*, 667–678. [CrossRef] [PubMed]
42. Beck, A.T. *Cognitive Therapy and the Emotional Disorders*; Meridian: New York, NY, USA, 1976; ISBN 978-1-101-65988-5.
43. Kendler, K.S.; Thornton, L.M.; Gardner, C.O. Stressful life events and previous episodes in the etiology of major depression in women: An evaluation of the "kindling" hypothesis. *Am. J. Psychiatry* **2000**, *157*, 1243–1251. [CrossRef] [PubMed]
44. American Psychiatric Association. *Diagnostic and Statistical Manual of Mental Disorders (DSM-5®)*; American Psychiatric Publishing: Washington, UK, 2013.
45. Atanasova, B.; Graux, J.; El Hage, W.; Hommet, C.; Camus, V.; Belzung, C. Olfaction: A potential cognitive marker of psychiatric disorders. *Neurosci. Biobehav. Rev.* **2008**, *32*, 1315–1325. [CrossRef] [PubMed]
46. Naudin, M.; El-Hage, W.; Gomes, M.; Gaillard, P.; Belzung, C.; Atanasova, B. State and trait olfactory markers of major depression. *PLoS ONE* **2012**, *7*, e46938. [CrossRef] [PubMed]
47. Naudin, M.; Carl, T.; Surguladze, S.; Guillen, C.; Gaillard, P.; Belzung, C.; El-Hage, W.; Atanasova, B. Perceptive Biases in Major Depressive Episode. *PLoS ONE* **2014**, *9*. [CrossRef] [PubMed]
48. Siopi, E.; Denizet, M.; Gabellec, M.-M.; de Chaumont, F.; Olivo-Marin, J.-C.; Guilloux, J.-P.; Lledo, P.-M.; Lazarini, F. Anxiety- and Depression-Like States Lead to Pronounced Olfactory Deficits and Impaired Adult Neurogenesis in Mice. *J. Neurosci. Off. J. Soc. Neurosci.* **2016**, *36*, 518–531. [CrossRef] [PubMed]
49. Pevsner, J.; Snyder, S.H. Odorant-binding protein: Odorant transport function in the vertebrate nasal epithelium. *Chem. Senses* **1990**, *15*, 217–222. [CrossRef]
50. Brand, G. *L'olfaction: De la Molécule au Comportement*; SOLAL Editeurs: Marseille, France, 2001; ISBN 978-2-914513-12-8.
51. Delplanque, S.; Coppin, G.; Sander, D. Odor and Emotion. In *Springer Handbook of Odor*; Buettner, A., Ed.; Springer International Publishing: Cham, Switzerland, 2017; pp. 101–102. ISBN 978-3-319-26930-6.
52. Parma, V.; Bulgheroni, M.; Tirindelli, R.; Castiello, U. Facilitation of action planning in children with autism: The contribution of the maternal body odor. *Brain Cogn.* **2014**, *88*, 73–82. [CrossRef] [PubMed]
53. Doty, R.L.; Shaman, P.; Dann, M. Development of the University of Pennsylvania Smell Identification Test: A standardized microencapsulated test of olfactory function. *Physiol. Behav.* **1984**, *32*, 489–502. [CrossRef]
54. Hummel, T.; Sekinger, B.; Wolf, S.R.; Pauli, E.; Kobal, G. "Sniffin' sticks': Olfactory performance assessed by the combined testing of odor identification, odor discrimination and olfactory threshold. *Chem. Senses* **1997**, *22*, 39–52. [CrossRef] [PubMed]
55. Doty, R.; Kobal, G. Current trends in the measurement of olfactory function. In *Handbook of Olfaction and Gustation*; John Wiley & Sons: New York, NY, USA, 1995; pp. 191–225.
56. Atanasova, B.; Hernandez, N.; van Nieuwenhuijzen, P.; Belzung, C. Psychophysical, neurophysiological and neurobiological investigation of the olfactory process in humans: Olfactory impairment in some neuropsychiatric disorders. In *The Biology of Odors: Sources, Olfaction and Response*; Nova Science Publishers Incorp: Hauppauge, NY, USA, 2011; pp. 1–67.
57. Brand, G.; Schaal, B. L'olfaction dans les troubles dépressifs: Intérêts et perspectives. *L'Encéphale* **2016**. [CrossRef] [PubMed]

58. Deems, D.A.; Doty, R.L.; Settle, R.G.; Moore-Gillon, V.; Shaman, P.; Mester, A.F.; Kimmelman, C.P.; Brightman, V.J.; Snow, J.B. Smell and taste disorders, a study of 750 patients from the University of Pennsylvania Smell and Taste Center. *Arch. Otolaryngol. Head Neck Surg.* **1991**, *117*, 519–528. [CrossRef] [PubMed]

59. Larsson, K. Impaired mating performances in male rats after anosmia induced peripherally or centrally. *Brain. Behav. Evol.* **1971**, *4*, 463–471. [CrossRef] [PubMed]

60. Kelly, J.P.; Leonard, B.E. Effects of chronic desipramine on waiting behaviour for a food reward in olfactory bulbectomized rats. *J. Psychopharmacol. (Oxf.)* **1996**, *10*, 153–156. [CrossRef] [PubMed]

61. Song, C.; Earley, B.; Leonard, B.E. The effects of central administration of neuropeptide Y on behavior, neurotransmitter, and immune functions in the olfactory bulbectomized rat model of depression. *Brain. Behav. Immun.* **1996**, *10*, 1–16. [CrossRef] [PubMed]

62. Kelly, J.P.; Wrynn, A.S.; Leonard, B.E. The olfactory bulbectomized rat as a model of depression: An update. *Pharmacol. Ther.* **1997**, *74*, 299–316. [CrossRef]

63. Negoias, S.; Croy, I.; Gerber, J.; Puschmann, S.; Petrowski, K.; Joraschky, P.; Hummel, T. Reduced olfactory bulb volume and olfactory sensitivity in patients with acute major depression. *Neuroscience* **2010**, *169*, 415–421. [CrossRef] [PubMed]

64. Li, Q.; Yang, D.; Wang, J.; Liu, L.; Feng, G.; Li, J.; Liao, J.; Wei, Y.; Li, Z. Reduced amount of olfactory receptor neurons in the rat model of depression. *Neurosci. Lett.* **2015**, *603*, 48–54. [CrossRef] [PubMed]

65. Kesner, R.P.; Gilbert, P.E.; Barua, L.A. The role of the hippocampus in memory for the temporal order of a sequence of odors. *Behav. Neurosci.* **2002**, *116*, 286–290. [CrossRef] [PubMed]

66. Lemogne, C.; Piolino, P.; Friszer, S.; Claret, A.; Girault, N.; Jouvent, R.; Allilaire, J.-F.; Fossati, P. Episodic autobiographical memory in depression: Specificity, autonoetic consciousness, and self-perspective. *Conscious. Cogn.* **2006**, *15*, 258–268. [CrossRef] [PubMed]

67. Pouliot, S.; Jones-Gotman, M. Medial temporal-lobe damage and memory for emotionally arousing odors. *Neuropsychologia* **2008**, *46*, 1124–1134. [CrossRef] [PubMed]

68. Drevets, W.C. Neuroimaging abnormalities in the amygdala in mood disorders. *Ann. N. Y. Acad. Sci.* **2003**, *985*, 420–444. [CrossRef] [PubMed]

69. Rogers, M.A.; Kasai, K.; Koji, M.; Fukuda, R.; Iwanami, A.; Nakagome, K.; Fukuda, M.; Kato, N. Executive and prefrontal dysfunction in unipolar depression: A review of neuropsychological and imaging evidence. *Neurosci. Res.* **2004**, *50*, 1–11. [CrossRef] [PubMed]

70. Grabenhorst, F.; Rolls, E.T.; Margot, C. A hedonically complex odor mixture produces an attentional capture effect in the brain. *NeuroImage* **2011**, *55*, 832–843. [CrossRef] [PubMed]

71. Zald, D.H.; Pardo, J.V. Emotion, olfaction, and the human amygdala: Amygdala activation during aversive olfactory stimulation. *Proc. Natl. Acad. Sci. USA* **1997**, *94*, 4119–4124. [CrossRef] [PubMed]

72. Grabenhorst, F.; Rolls, E.T.; Margot, C.; da Silva, M.A.A.P.; Velazco, M.I. How pleasant and unpleasant stimuli combine in different brain regions: Odor mixtures. *J. Neurosci. Off. J. Soc. Neurosci.* **2007**, *27*, 13532–13540. [CrossRef] [PubMed]

73. Zald, D.H.; Mattson, D.L.; Pardo, J.V. Brain activity in ventromedial prefrontal cortex correlates with individual differences in negative affect. *Proc. Natl. Acad. Sci. USA* **2002**, *99*, 2450–2454. [CrossRef] [PubMed]

74. Phillips, M.L.; Drevets, W.C.; Rauch, S.L.; Lane, R. Neurobiology of emotion perception II: Implications for major psychiatric disorders. *Biol. Psychiatry* **2003**, *54*, 515–528. [CrossRef]

75. Bhagwagar, Z.; Wylezinska, M.; Jezzard, P.; Evans, J.; Boorman, E.; Matthews, P.M.; Cowen, P.J. Low GABA concentrations in occipital cortex and anterior cingulate cortex in medication-free, recovered depressed patients. *Int. J. Neuropsychopharmacol.* **2008**, *11*, 255–260. [CrossRef] [PubMed]

76. Fulbright, R.K.; Skudlarski, P.; Lacadie, C.M.; Warrenburg, S.; Bowers, A.A.; Gore, J.C.; Wexler, B.E. Functional MR imaging of regional brain responses to pleasant and unpleasant odors. *AJNR Am. J. Neuroradiol.* **1998**, *19*, 1721–1726. [PubMed]

77. Paulus, M.P.; Stein, M.B. An insular view of anxiety. *Biol. Psychiatry* **2006**, *60*, 383–387. [CrossRef] [PubMed]

78. Veldhuizen, M.G.; Nachtigal, D.; Teulings, L.; Gitelman, D.R.; Small, D.M. The Insular Taste Cortex Contributes to Odor Quality Coding. *Front. Hum. Neurosci.* **2010**, *4*. [CrossRef] [PubMed]

79. Sliz, D.; Hayley, S. Major Depressive Disorder and Alterations in Insular Cortical Activity: A Review of Current Functional Magnetic Imaging Research. *Front. Hum. Neurosci.* **2012**, *6*. [CrossRef] [PubMed]

80. Shumake, J.; Edwards, E.; Gonzalez-Lima, F. Opposite metabolic changes in the habenula and ventral tegmental area of a genetic model of helpless behavior. *Brain Res.* **2003**, *963*, 274–281. [CrossRef]

81. Da Costa, A.P.; Broad, K.D.; Kendrick, K.M. Olfactory memory and maternal behaviour-induced changes in c-fos and zif/268 mRNA expression in the sheep brain. *Brain Res. Mol. Brain Res.* **1997**, *46*, 63–76. [CrossRef]

82. Hikosaka, O.; Sesack, S.R.; Lecourtier, L.; Shepard, P.D. Habenula—Crossroad between the Basal Ganglia and the Limbic System. *J. Neurosci. Off. J. Soc. Neurosci.* **2008**, *28*, 11825–11829. [CrossRef] [PubMed]

83. Oral, E.; Aydin, M.D.; Aydin, N.; Ozcan, H.; Hacimuftuoglu, A.; Sipal, S.; Demirci, E. How olfaction disorders can cause depression? The role of habenular degeneration. *Neuroscience* **2013**, *240*, 63–69. [CrossRef] [PubMed]

84. Pause, B.M.; Raack, N.; Sojka, B.; Göder, R.; Aldenhoff, J.B.; Ferstl, R. Convergent and divergent effects of odors and emotions in depression. *Psychophysiology* **2003**, *40*, 209–225. [CrossRef] [PubMed]

85. Lombion-Pouthier, S.; Vandel, P.; Nezelof, S.; Haffen, E.; Millot, J.-L. Odor perception in patients with mood disorders. *J. Affect. Disord.* **2006**, *90*, 187–191. [CrossRef] [PubMed]

86. Pause, B.M.; Miranda, A.; Göder, R.; Aldenhoff, J.B.; Ferstl, R. Reduced olfactory performance in patients with major depression. *J. Psychiatr. Res.* **2001**, *35*, 271–277. [CrossRef]

87. Postolache, T.T.; Doty, R.L.; Wehr, T.A.; Jimma, L.A.; Han, L.; Turner, E.H.; Matthews, J.R.; Neumeister, A.; No, C.; Kroger, H.; et al. Monorhinal odor identification and depression scores in patients with seasonal affective disorder. *J. Affect. Disord.* **1999**, *56*, 27–35. [CrossRef]

88. Scinska, A.; Wrobel, E.; Korkosz, A.; Zatorski, P.; Sienkiewicz-Jarosz, H.; Lojkowska, W.; Swiecicki, L.; Kukwa, W. Depressive symptoms and olfactory function in older adults. *Psychiatry Clin. Neurosci.* **2008**, *62*, 450–456. [CrossRef] [PubMed]

89. Swiecicki, L.; Zatorski, P.; Bzinkowska, D.; Sienkiewicz-Jarosz, H.; Szyndler, J.; Scinska, A. Gustatory and olfactory function in patients with unipolar and bipolar depression. *Prog. Neuropsychopharmacol. Biol. Psychiatry* **2009**, *33*, 827–834. [CrossRef] [PubMed]

90. Gross-Isseroff, R.; Luca-Haimovici, K.; Sasson, Y.; Kindler, S.; Kotler, M.; Zohar, J. Olfactory sensitivity in major depressive disorder and obsessive compulsive disorder. *Biol. Psychiatry* **1994**, *35*, 798–802. [CrossRef]

91. Pollatos, O.; Albrecht, J.; Kopietz, R.; Linn, J.; Schoepf, V.; Kleemann, A.M.; Schreder, T.; Schandry, R.; Wiesmann, M. Reduced olfactory sensitivity in subjects with depressive symptoms. *J. Affect. Disord.* **2007**, *102*, 101–108. [CrossRef] [PubMed]

92. Clepce, M.; Gossler, A.; Reich, K.; Kornhuber, J.; Thuerauf, N. The relation between depression, anhedonia and olfactory hedonic estimates—A pilot study in major depression. *Neurosci. Lett.* **2010**, *471*, 139–143. [CrossRef] [PubMed]

93. Serby, M.; Larson, P.; Kalkstein, D. Olfactory sense in psychoses. *Biol. Psychiatry* **1990**, *28*, 830. [CrossRef]

94. Zucco, G.M.; Bollini, F. Odour recognition memory and odour identification in patients with mild and severe major depressive disorders. *Psychiatry Res.* **2011**, *190*, 217–220. [CrossRef] [PubMed]

95. Atanasova, B.; El-Hage, W.; Chabanet, C.; Gaillard, P.; Belzung, C.; Camus, V. Olfactory anhedonia and negative olfactory alliesthesia in depressed patients. *Psychiatry Res.* **2010**, *176*, 190–196. [CrossRef] [PubMed]

96. Amsterdam, J.D.; Settle, R.G.; Doty, R.L.; Abelman, E.; Winokur, A. Taste and smell perception in depression. *Biol. Psychiatry* **1987**, *22*, 1481–1485. [CrossRef]

97. Kopala, L.C.; Good, K.P.; Honer, W.G. Olfactory hallucinations and olfactory identification ability in patients with schizophrenia and other psychiatric disorders. *Schizophr. Res.* **1994**, *12*, 205–211. [CrossRef]

98. McCaffrey, R.J.; Duff, K.; Solomon, G.S. Olfactory dysfunction discriminates probable Alzheimer's dementia from major depression: A cross-validation and extension. *J. Neuropsychiatry Clin. Neurosci.* **2000**, *12*, 29–33. [CrossRef] [PubMed]

99. Pentzek, M.; Grass-Kapanke, B.; Ihl, R. Odor identification in Alzheimer's disease and depression. *Aging Clin. Exp. Res.* **2007**, *19*, 255–258. [CrossRef] [PubMed]

100. Solomon, G.S.; Petrie, W.M.; Hart, J.R.; Brackin, J.; Henry, B. Olfactory Dysfunction Discriminates Alzheimer's Dementia from Major Depression. *J. Neuropsychiatry Clin. Neurosci.* **1998**, *10*, 64–67. [CrossRef] [PubMed]

101. Atanasova, B.; Gaillard, P.; Minier, F.; Belzung, C.; El-Hage, W. Hedonic Olfactory Perception in Depression: Relationship between Self-Evaluation and Autonomic Response. *Psychology* **2012**, *3*, 959–965. [CrossRef]

102. Naudin, M.; Mondon, K.; El-Hage, W.; Desmidt, T.; Jaafari, N.; Belzung, C.; Gaillard, P.; Hommet, C.; Atanasova, B. Long-term odor recognition memory in unipolar major depression and Alzheimer's disease. *Psychiatry Res.* **2014**, *220*, 861–866. [CrossRef] [PubMed]

103. Thomas, H.J.; Fries, W.; Distel, H. Evaluation of olfactory stimuli by depressed patients. *Nervenarzt* **2002**, *73*, 71–77. [CrossRef] [PubMed]

104. Postolache, T.T.; Wehr, T.A.; Doty, R.L.; Sher, L.; Turner, E.H.; Bartko, J.J.; Rosenthal, N.E. Patients with Seasonal Affective Disorder Have Lower Odor Detection Thresholds Than Control Subjects. *Arch. Gen. Psychiatry* **2002**, *59*, 1119–1122. [CrossRef] [PubMed]

105. Oren, D.A.; Schwartz, P.J.; Turner, E.H.; Rosenthal, N.E. Olfactory function in winter seasonal affective disorder. *Am. J. Psychiatry* **1995**, *152*, 1531–1532. [PubMed]

106. Kamath, V.; Paksarian, D.; Cui, L.; Moberg, P.J.; Turetsky, B.I.; Merikangas, K.R. Olfactory processing in bipolar disorder, major depression, and anxiety. *Bipolar Disord.* **2018**. [CrossRef] [PubMed]

107. Striebel, K.M.; Beyerstein, B.; Remick, R.A.; Kopala, L.; Honer, W.G. Olfactory identification and psychosis. *Biol. Psychiatry* **1999**, *45*, 1419–1425. [CrossRef]

108. Krüger, S.; Frasnelli, J.; Bräunig, P.; Hummel, T. Increased olfactory sensitivity in euthymic patients with bipolar disorder with event-related episodes compared with patients with bipolar disorder without such episodes. *J. Psychiatry Neurosci.* **2006**, *31*, 263–270. [PubMed]

109. Vasterling, J.J.; Brailey, K.; Sutker, P.B. Olfactory identification in combat-related posttraumatic stress disorder. *J. Trauma. Stress* **2000**, *13*, 241–253. [CrossRef] [PubMed]

110. McDowell, I. *Measuring Health*; Oxford University Press: Oxford, UK, 2006; ISBN 978-0-19-516567-8.

111. Derogatis, L.R.; Unger, R. Symptom Checklist-90-Revised. In *The Corsini Encyclopedia of Psychology*; Weiner, I.B., Craighead, W.E., Eds.; John Wiley & Sons, Inc.: Hoboken, NJ, USA, 2010; ISBN 978-0-470-47921-6.

112. Wiklund, I.; Romanus, B.; Hunt, S.M. Self-assessed disability in patients with arthrosis of the hip joint. Reliability of the Swedish version of the Nottingham Health Profile. *Int. Disabil. Stud.* **1988**, *10*, 159–163. [CrossRef] [PubMed]

113. McDowell, I. Measures of self-perceived well-being. *J. Psychosom. Res.* **2010**, *69*, 69–79. [CrossRef] [PubMed]

114. Temmel, A.F.P.; Quint, C.; Schickinger-Fischer, B.; Klimek, L.; Stoller, E.; Hummel, T. Characteristics of olfactory disorders in relation to major causes of olfactory loss. *Arch. Otolaryngol. Head Neck Surg.* **2002**, *128*, 635–641. [CrossRef] [PubMed]

115. Ferris, A.M.; Duffy, V.B. Effect of olfactory deficits on nutritional status. Does age predict persons at risk? *Ann. N. Y. Acad. Sci.* **1989**, *561*, 113–123. [CrossRef] [PubMed]

116. Blomqvist, E.H.; Brämerson, A.; Stjärne, P.; Nordin, S. Consequences of olfactory loss and adopted coping strategies. *Rhinology* **2004**, *42*, 189–194. [PubMed]

117. Nordin, S.; Hedén Blomqvist, E.; Olsson, P.; Stjärne, P.; Ehnhage, A.; NAF2S2 Study Group. Effects of smell loss on daily life and adopted coping strategies in patients with nasal polyposis with asthma. *Acta Otolaryngol. (Stockh.)* **2011**, *131*, 826–832. [CrossRef] [PubMed]

118. Nordin, S.; Brämerson, A.; Murphy, C.; Bende, M. A Scandinavian adaptation of the Multi-Clinic Smell and Taste Questionnaire: Evaluation of questions about olfaction. *Acta Otolaryngol. (Stockh.)* **2003**, *123*, 536–542. [CrossRef]

119. Croy, I.; Negoias, S.; Novakova, L.; Landis, B.N.; Hummel, T. Learning about the functions of the olfactory system from people without a sense of smell. *PLoS ONE* **2012**, *7*, e33365. [CrossRef] [PubMed]

120. Santos, D.V.; Reiter, E.R.; DiNardo, L.J.; Costanzo, R.M. Hazardous events associated with impaired olfactory function. *Arch. Otolaryngol. Head Neck Surg.* **2004**, *130*, 317–319. [CrossRef] [PubMed]

121. Brämerson, A.; Johansson, L.; Ek, L.; Nordin, S.; Bende, M. Prevalence of olfactory dysfunction: The skövde population-based study. *Laryngoscope* **2004**, *114*, 733–737. [CrossRef] [PubMed]

122. Ackerl, K.; Atzmueller, M.; Grammer, K. The scent of fear. *Neuro Endocrinol. Lett.* **2002**, *23*, 79–84. [PubMed]

123. Prehn-Kristensen, A.; Wiesner, C.; Bergmann, T.O.; Wolff, S.; Jansen, O.; Mehdorn, H.M.; Ferstl, R.; Pause, B.M. Induction of empathy by the smell of anxiety. *PLoS ONE* **2009**, *4*, e5987. [CrossRef] [PubMed]

124. Schaal, B.; Montagner, H.; Hertling, E.; Bolzoni, D.; Moyse, A.; Quichon, R. Les stimulations olfactives dans les relations entre l'enfant et la mère. *Reprod. Nutr. Dév.* **1980**, *20*, 843–858. [CrossRef] [PubMed]

125. Gelstein, S.; Yeshurun, Y.; Rozenkrantz, L.; Shushan, S.; Frumin, I.; Roth, Y.; Sobel, N. Human tears contain a chemosignal. *Science* **2011**, *331*, 226–230. [CrossRef] [PubMed]

126. Stevenson, R.J. An initial evaluation of the functions of human olfaction. *Chem. Senses* **2010**, *35*, 3–20. [CrossRef] [PubMed]

127. Komori, T.; Fujiwara, R.; Tanida, M.; Nomura, J. Potential antidepressant effects of lemon odor in rats. *Eur. Neuropsychopharmacol. J. Eur. Coll. Neuropsychopharmacol.* **1995**, *5*, 477–480. [CrossRef]

128. Harkin, A.; Kelly, J.P.; Leonard, B.E. A review of the relevance and validity of olfactory bulbectomy as a model of depression. *Clin. Neurosci. Res.* **2003**, *3*, 253–262. [CrossRef]

129. Lehrner, J.; Eckersberger, C.; Walla, P.; Pötsch, G.; Deecke, L. Ambient odor of orange in a dental office reduces anxiety and improves mood in female patients. *Physiol. Behav.* **2000**, *71*, 83–86. [CrossRef]

130. Diego, M.A.; Jones, N.A.; Field, T.; Hernandez-Reif, M.; Schanberg, S.; Kuhn, C.; McAdam, V.; Galamaga, R.; Galamaga, M. Aromatherapy positively affects mood, EEG patterns of alertness and math computations. *Int. J. Neurosci.* **1998**, *96*, 217–224. [CrossRef] [PubMed]

131. Lehrner, J.; Marwinski, G.; Lehr, S.; Johren, P.; Deecke, L. Ambient odors of orange and lavender reduce anxiety and improve mood in a dental office. *Physiol. Behav.* **2005**, *86*, 92–95. [CrossRef] [PubMed]

132. Lis-Balchin, M.; Hart, S. Studies on the mode of action of the essential oil of Lavender (*Lavandula angustifolia* P. Miller). *Phytother. Res.* **1999**, *13*, 540–542. [CrossRef]

133. Herz, R.S. Aromatherapy facts and fictions: A scientific analysis of olfactory effects on mood, physiology and behavior. *Int. J. Neurosci.* **2009**, *119*, 263–290. [CrossRef] [PubMed]

134. Maheswaran, T.; Abikshyeet, P.; Sitra, G.; Gokulanathan, S.; Vaithiyanadane, V.; Jeelani, S. Gustatory dysfunction. *J. Pharm. Bioallied Sci.* **2014**, *6*, S30–S33. [CrossRef] [PubMed]

135. Ribeiro, M.V.M. R.; Hasten-Reiter Júnior, H.N.; Ribeiro, E.A.N.; Jucá, M.J.; Barbosa, F.T.; Sousa-Rodrigues, C.F. de Association between visual impairment and depression in the elderly: A systematic review. *Arq. Bras. Oftalmol.* **2015**, *78*, 197–201. [CrossRef] [PubMed]

136. Mosaku, K.; Akinpelu, V.; Ogunniyi, G. Psychopathology among a sample of hearing impaired adolescents. *Asian J. Psychiatry* **2015**, *18*, 53–56. [CrossRef] [PubMed]

Review

Transcranial Direct Current Stimulation (tDCS): A Promising Treatment for Major Depressive Disorder?

Djamila Bennabi * and Emmanuel Haffen

Department of Clinical Psychiatry, CIC-1431 INSERM, CHU de Besançon, EA Neurosciences, University Bourgogne Franche-Comte, FondaMental Foundation, 94000 Creteil, France; emmanuel.haffen@univ-fcomte.fr
* Correspondence: djamila.bennabi@univ-fcomte.fr; Tel.: +33-381-218-454

Received: 7 March 2018; Accepted: 3 May 2018; Published: 6 May 2018

Abstract: Background: Transcranial direct current stimulation (tDCS) opens new perspectives in the treatment of major depressive disorder (MDD), because of its ability to modulate cortical excitability and induce long-lasting effects. The aim of this review is to summarize the current status of knowledge regarding tDCS application in MDD. **Methods:** In this review, we searched for articles published in PubMed/MEDLINE from the earliest available date to February 2018 that explored clinical and cognitive effects of tDCS in MDD. **Results:** Despite differences in design and stimulation parameters, the examined studies indicated beneficial effects of tDCS for MDD. These preliminary results, the non-invasiveness of tDCS, and its good tolerability support the need for further research on this technique. **Conclusions:** tDCS constitutes a promising therapeutic alternative for patients with MDD, but its place in the therapeutic armamentarium remains to be determined.

Keywords: transcranial direct current stimulation; depression; cognition

1. Introduction

Major depressive disorder (MDD) is a widespread psychiatric disease characterized by a high risk of morbidity and mortality and a high level of comorbidity with several psychiatric and non-psychiatric disorders [1]. Epidemiological surveys have repeatedly indicated a high lifetime prevalence of this illness, amounting to 6.7% of the population worldwide [2]. The costs of MDD to society, in terms of personal and familial suffering and health care consumption, are high [3,4]. Despite the progress in pharmacopoeia and in psychological therapies, clinicians involved in the management of MDD are regularly faced with clinical situations marked by treatment resistance. About 30% of depressed patients fail to experience significant clinical benefits from currently available treatments [5,6], leading to a chronically deteriorating course of the illness. Consequences of the illness and limitations of the usual pharmacological and psychological strategies highlight the necessity to develop alternative therapeutic options.

Brain stimulation therapies have emerged as relevant alternative strategies, on the basis of emerging knowledge about specific brain areas involved in psychiatric diseases. Among these techniques, transcranial direct current stimulation (tDCS) appears to hold particular promise because of its cost, ease of use, and favorable tolerability profile [7]. tDCS was used from the 1960s to generate modifications of cortical excitability in preclinical studies and as a therapeutic tool for MDD. More recently, this technique has gained renewed interest as a practical tool for the modulation of cortical excitability and the treatment of psychiatric disorders [7,8]. tDCS relies on the application of a weak direct current of 1–2 mA directly to the scalp through electrodes to induce regional changes in cortical excitability that can last up to a few hours after stimulation [9]. Sustainable effects seem to be mediated by bidirectional modifications of postsynaptic connections similar to long-term potentiation and depression, occurring through NMDA-dependent mechanisms [10,11]. Because of the implication of pathological alterations of neuroplasticity in psychiatric disorders, tDCS appears to be a promising therapeutic alternative to modify such pathological plasticity [12]. Beneficial effects of tDCS have been reported in the treatment of psychiatric (mainly depression and schizophrenia) and neurological diseases, as well as in the rehabilitation of cognitive, motor, and sensory functions after a stroke [8,13].

In this review, we summarize data obtained from trials with the aim of assessing the effectiveness of tDCS as a putative treatment option for MDD.

1.1. Method

To identify studies reporting the effects of tDCS on depression and cognition in MDD, two authors (E.H. and D.B.) searched the Pub-med database following PRISMA recommendations [14]. The identification of articles was based on the following keywords: "depression", "transcranial direct current stimulation".

Inclusion criteria for this review were: (a) publication in English; (b) inclusion of depressed patients treated with tDCS; (c) meta-analysis, randomized controlled studies (RCTs) and open-label trials; (d) assessment of depressive symptoms and/or cognition. Working independently and in duplicate, the two reviewers examined all titles and abstracts, obtained full texts of potentially relevant papers, and read the papers to determine whether they met the inclusion criteria. Of the 381 initial references, 67 papers were retained. We excluded seven studies exploring the effects of tDCS in healthy subjects, one review, and one protocol. Thus, we obtained data from 58 articles that met our eligibility criteria.

1.2. Technical and Safety Aspects of Transcranial Direct Current Stimulation

During tDCS sessions, a constant direct current of low intensity (generally 1–2 mA) is passed across the brain via electrodes wrapped in an electrode gel or saline-soaked sponge pockets. [8]. Wide variations in the amount of current delivered can be observed according to the stimulation parameters applied (i.e., position and size of the electrodes, current intensity, duration, frequency, and number of sessions). In the majority of protocols, the currents ranged from 0.5 to 2 mA and were delivered for 5–30 min via electrodes of 25–35 cm, generating current densities of 0.28–0.8 mA/cm^2. The placement of the electrodes is typically based on the 10–20 electrode placement system to locate the area of interest. More recent studies suggest the use of an initial magnetic resonance imaging (MRI) scan to refine the position of the electrode, taking into account the inter-individual variability [15]. From a neurobiological standpoint, tDCS is considered a technique of neuromodulation, given that it does not directly induce action potentials, unlike transcranial magnetic stimulation. The polarity of the stimulation determines the type of effect; anodal stimulation induces a depolarization of the neuronal membranes and an increase of the spontaneous neuronal firing rate, whereas cathodal stimulation leads to neuronal hyperpolarization and inhibition [16,17]. Depending on the duration and intensity of the stimulation, it has been shown that the modulation of electrophysiological properties is directly measurable in healthy volunteers and could last up to 90 min after stopping the stimulation [9].

tDCS may induce mild to moderate side effects, including light itching beneath the electrodes, mild headache, tingling, burning sensations, and discomfort. Skin irritation and lesions under the electrodes have also been reported, as well as some cases of mood switching [18,19].

1.3. Effects of tDCS on Symptoms

The rationale for using tDCS in this indication is based on historical observations of hypometabolism of the left dorsolateral prefrontal cortex (dlPFC) associated with right prefrontal hyperfunction in MDD and dysfunction of brain plasticity, characterized by an alteration of long-term potentiation and depression [20,21]. In the majority of the protocols, the currents used were 1 or 2 mA and were applied for 30 min via two large conductive electrodes (32–35 cm^2) soaked with a saline solution. The anode was typically placed over the left dlPFC, and the cathode over the contralateral supraorbital area, corresponding to F3 and FP2 according to the international 10–20 EEG system [22]. Transcranial direct current stimulation was proposed either alone or as an add-on treatment to psychotropic medication or cognitive training programs.

Following the seminal work of Fregni et al. [23] which produced positive results on the efficacy of five sessions of anodal tDCS applied over the left dlPFC (1 mA, 10 min) in ten patients with MDD, a series of controlled and open-label studies have been published. Subsequently, two RCTs conducted by Boggio et al. [24,25] have replicated these results. The authors observed an average reduction of 40.4% in the Hamilton Depression Rating Scale (HDRS) score after anodal tDCS (2 mA, 20 min) versus 10.4% after placebo stimulation in 40 patients with mild to moderate MDD. In 2010, Loo et al. [26] showed no superior antidepressant efficacy of ten sessions of tDCS (1 mA, 20 min) versus placebo in 40 patients with MDD. However, in a second study, these authors demonstrated a greater decrease in the Hamilton Depression Rating Scale (HDRS scores in 64 unipolar and bipolar depressed patients after 15 sessions of anodal tDCS of the left dlPFC (2 mA, 20 min) (28.4%) versus placebo (15.9%) [27]. Rigonatti et al. [28] compared the effect of fluoxetine and ten tDCS sessions (2 mA, 20 min) in 42 depressed patients, and noted a similar improvement in depressive symptoms following brain stimulation and pharmacological treatment, with an earlier antidepressant action in the tDCS group. In the field of treatment-resistant depression, three RCTs were conducted and did not show any significant difference between active stimulation and placebo [29–31]. Regarding post-stroke depression, Valiengo et al. [32] studied the effects after active versus sham stimulation in 45 patients and reported a higher response rate with active stimulation (37.5% and 20.8%, respectively) and higher remission rates (4.1% and 0%, respectively).

In this context, in 2013, a larger study was conducted involving 120 unipolar depressed patients, which compared tDCS versus a pharmacological treatment (sertraline) and versus tDCS plus sertraline. The results showed a greater reduction in Montgomery–Asberg Depression Rating Scale scores in patients receiving the combined intervention (tDCS + sertraline) versus those receiving sertraline alone (mean difference 8.5 points), tDCS alone (mean difference 5.9 points), or placebo (mean difference 11.5 points) [33]. More recently, these authors published the results of a non-inferiority trial involving 245 depressed patients and compared active tDCS (2 mA, 30 min) versus a pharmacological treatment (escitalopram) and versus placebo. They concluded that escitalopram was significantly superior to tDCS [34]. Moreover, a large-scale RCT involving 84 unipolar and 36 bipolar depressed patients reported comparable effects of active (2.5 mA, 30 min, 20 sessions over four weeks) and sham tDCS applied over the left dlPFC, suggesting that the low level of stimulation delivered in sham conditions may have been biologically active [35].

Studies evaluating the effectiveness of tDCS in a maintenance therapy for relapse prevention have shown that the reduction in treatment frequency from two to one week or a high level of pre-treatment resistance was associated with a greater rate of relapse [36,37].

Several meta-analyses pooling available data about the antidepressant efficacy of tDCS have yielded contradictory results. Kalu et al. [38] showed a higher efficacy of active tDCS versus placebo, as evidenced by a greater reduction in severity scores on depression scales. Berlim et al. [39] did not find any significant difference between active tDCS and placebo in response rates (23.3% versus 12.4%, respectively, $p = 0.11$) and remission (12.2% versus 5.4%, respectively, $p = 0.22$). Shiozawa et al. [40] included seven controlled trials in their meta-analysis and showed a superiority of active tDCS over placebo in terms of clinical response and remission, which was confirmed by Meron et al. [41] in a meta-analysis of 10 RCTs. In a meta-analysis of individual data from 289 patients, Brunoni et al. [42] demonstrated the superiority of active tDCS compared to placebo in terms of alleviation of depressive symptoms, with a response rate of 33.3% versus 19 %, respectively, and a remission rate of 23.1 versus 12.7%, respectively. The level of response was correlated with various parameters, namely, the number of sessions and the amount of energy delivered, but was inversely associated with the level of antidepressant resistance. Other variables, such as the severity of the current depressive episode, the presence of bipolar disorder, female gender, or treatment with sertraline, as well as pre-treatment motor retardation or better verbal fluency, were also identified as potential predictors of a positive response [42–45]. From a neurobiological point of view, 5-HTTLPR polymorphism has shown a predictive property, while brain-derived neurotrophic factor, cytokine levels, neurotrophins 3 and 4, nerve growth factor, and glial cell line-derived neurotropic factor have failed to predict the clinical response [46–49].

1.4. Effects of tDCS on Cognition

Current evidence suggests that tDCS may have an impact on some cognitive functions in healthy volunteers, such as working memory, attentional performance, procedural learning, and emotional information processing [50]. In MDD, although most studies reported an improvement in at least some cognitive functions, suggesting a potential pro-cognitive role of tDCS, no firm conclusions could be drawn. To date, the improvement of attention and working memory has been reported after 1 [27,51,52], 5 [53], and 10 [25] anodal tDCS sessions applied over the left DLPFC. Positive results have also been observed in other cognitive domains, such as cognitive control [54,55], processing speed [56], or emotion recognition [57]. Bifrontal tDCS has been shown to promote more accurate and faster responses to the n-back task, exploring working memory, and to prevent procedural learning during the probabilistic classification learning task in depressive states [51]. Moreover, a few clinical cases of improvement in cognitive performances after treatment with tDCS have been reported in the context of treatment-resistant depression or post-traumatic depression [29,58]. However, several RCTs applying a set of standardized cognitive tests did not record tDCS-related changes with offline stimulations, suggesting that repeated sessions have no cumulative effects [26,31,59]. Beyond the variability in the stimulations parameters (i.e., site of stimulation or use of off- or online sessions) and the impact of inter- and intraindividual differences, in most of the studies it is difficult to differentiate the "pure" effects of cognitive improvement of tDCS from a "collateral" effect relied on a cognitive improvement due to the alleviation of depression [60].

1.5. Parameters Influencing Clinical Outcomes

Multiple factors are likely to modulate the therapeutic effects of tDCS. Among them, the stimulation parameters and clinical characteristics of the population are key contributors to the variability of its effects [60–62]. With regard to MDD, there is a dearth of clinical trials exploring the impact of the stimulation parameters on clinical outcomes. Typically, bifrontal montages (F3–F8 and F3–F4 montages) targeting the left dlPFC are used in MDD. However, a computational modelling study suggested that other montages, using a fronto-extracephalic or fronto-occipital approach, could result in greater stimulation of brain regions of critical interest, such as the anterior cingulate cortex, which may be advantageous for treating MDD [63]. In fact, two open-label trials observed improvement in depressive symptoms following tDCS sessions applied over the fronto-occipital or -temporal regions [64,65] in a total of 18 patients and 20 sessions. Moreover, combined with sertraline, tDCS applied for 20 or 30 min was found to be effective for the treatment of mild and moderate depression, and the effect of the stimulation for 30 min was more pronounced than that of the 20 min stimulation [66]. Meta-analyses have noted that increasing the number of sessions or the intensity of the stimulation (1 versus 2 mA) does not improve tDCS effects on depressive symptoms [38,39]. In addition, the delay between sessions could have an impact on tDCS effects [67]. For example, Alonzo et al. [68] reported that daily tDCS induced a greater increase in cortical excitability than second daily stimulation of the motor cortex. Another, often overlooked, point is the influence of patient characteristics. Two open-label trials reported that depression severity was positively related to clinical improvement [69,70]. Evidence from three RCTs indicated that dlPFC tDCS had lesser efficacy in treatment-resistant depressed patients [29,31]. Finally, a meta-analysis of data from seven RCTs showed the superiority of active tDCS versus placebo in bipolar depression in terms of reduction in severity scores on depression scales from the first week of treatment [71]. Besides the features of the ongoing episode, the effects of interindividual variables, such as anatomical differences, genetic factors, personality, comorbidities, cognitive strategy, lifestyle, and baseline neuronal activation state, need to be explored [50]. Likewise, the age at which the stimulation is delivered might be a key determinant of the physiological and behavioral outcomes of tDCS. In children and adolescents, specific effects of tDCS on cortical excitability have been demonstrated [72,73]. In this population, the choice of the stimulation parameters, especially the dose selection, requires special attention [74].

It should also be emphasized that the final effects of tDCS depend on the concomitant use of pharmacotherapy with other interventions, such as cognitive therapy. A synergistic therapeutic action was observed with the combination of tDCS and sertraline in MDD patients in a large-scale RCT which compared the efficacy of tDCS, sertraline, and a combination of both [33]. Conversely, benzodiazepines were reported to reduce tDCS effects [75]. Concerning the effects of adjunctive tDCS and cognitive control therapy, Segrave et al. [76] showed that active tDCS coupled with weekly cognitive behavioral therapy (CBT) potentiated the treatment response, while Brunoni et al. [77] failed to demonstrate the superiority of combined cognitive control training (CCT) and active tDCS intervention versus CCT and sham tDCS. In resistant depression, one open study reported an improvement in depressive symptoms after 18 sessions of tDCS over six weeks, administered during Cognitive Emotional Training [78]. More recently, Mayur et al. failed to demonstrate differences in terms of speed of response or cognitive performance after the use of tDCS in combination with electroconvulsive therapy (ECT) versus ECT alone [79].

2. Conclusions

tDCS is a promising therapeutic strategy that offers the opportunity for non-invasive modulation of cortical excitability and plasticity in psychiatric disorders. With regard to MDD, the majority of meta-analyses have found that tDCS is superior to sham stimulation with an effect size (B coefficient = 0.35) comparable to that of repetitive transcranial magnetic stimulation (rTMS) and antidepressant medication in primary care [42]. This technique appears to be particularly indicated for patients with a mild-to-severe form of MDD who do not meet the criteria for resistant depression. In line with these data, a European expert group has recently proposed a Level B recommendation for its use in depressed patients, treated or not with antidepressants, without treatment resistance (minimum of 10 sessions (2 mA, 20–30 min) with the anode over the left dlPFC and the cathode over the right supra-orbital region) [7]. Questions still remain unanswered regarding the optimal stimulation parameters, the effect of tasks given during tDCS sessions, and the possible influence of add-on medications. Moreover, the clinical profile of depressed patients showing favorable responses to tDCS require clarification. Besides these critical questions, the promising preliminary results, the non-invasiveness of tDCS, and its good tolerability support the need for further research into this technique.

Author Contributions: Both authors contributed substantially to the preparation of the manuscript and approved the submitted version.

Acknowledgments: The authors received no specific funding for this work.

Conflicts of Interest: The authors declare no conflict of interest.

References

1.   Rush, A.J.; Zimmerman, M.; Wisniewski, S.R.; Fava, M.; Hollon, S.D.; Warden, D.; Biggs, M.M.; Shores-Wilson, K.; Shelton, R.C.; Luther, J.F.; et al. Comorbid psychiatric disorders in depressed outpatients: Demographic and clinical features. *J. Affect. Disord.* **2005**, *87*, 43–55. [CrossRef] [PubMed]

2.   Wittchen, H.U.; Jacobi, F.; Rehm, J.; Gustavsson, A.; Svensson, M.; Jönsson, B.; Olesen, J.; Allgulander, C.; Alonso, J.; Faravelli, C.; et al. The size and burden of mental disorders and other disorders of the brain in Europe 2010. *Eur. Neuropsychopharmacol.* **2011**, *21*, 655–679. [CrossRef] [PubMed]

3.   Olesen, J.; Gustavsson, A.; Svensson, M.; Wittchen, H.-U.; Jönsson, B.; CDBE2010 Study Group. European Brain Council. The economic cost of brain disorders in Europe. *Eur. J. Neurol.* **2012**, *19*, 155–162. [CrossRef] [PubMed]

4.   McCrone, P.; Rost, F.; Koeser, L.; Koutoufa, I.; Stephanou, S.; Knapp, M.; Goldberg, D.; Taylor, D.; Fonagy, P. The economic cost of treatment-resistant depression in patients referred to a specialist service. *J. Ment. Health* **2017**, 1–7. [CrossRef] [PubMed]

5.   Rush, A.J.; Trivedi, M.H.; Wisniewski, S.R.; Nierenberg, A.A.; Stewart, J.W.; Warden, D.; Niederehe, G.; Thase, M.E.; Lavori, P.W.; Lebowitz, B.D.; et al. Acute and longer-term outcomes in depressed outpatients requiring one or several treatment steps: A STAR*D report. *Am. J. Psychiatry* **2006**, *163*, 1905–1917. [CrossRef] [PubMed]

6.   Trivedi, M.H.; Rush, A.J.; Wisniewski, S.R.; Nierenberg, A.A.; Warden, D.; Ritz, L.; Norquist, G.; Howland, R.H.; Lebowitz, B.; McGrath, P.J.; et al. STAR*D study team evaluation of outcomes with citalopram for depression using measurement-based care in STAR*D: Implications for clinical practice. *Am. J. Psychiatry* **2006**, *163*, 28–40. [CrossRef] [PubMed]

7.   Lefaucheur, J.-P.; Antal, A.; Ayache, S.S.; Benninger, D.H.; Brunelin, J.; Cogiamanian, F.; Cotelli, M.; De Ridder, D.; Ferrucci, R.; Langguth, B.; et al. Evidence-based guidelines on the therapeutic use of transcranial direct current stimulation (tDCS). *Clin. Neurophysiol.* **2017**, *128*, 56–92. [CrossRef] [PubMed]

8.   Mondino, M.; Bennabi, D.; Poulet, E.; Galvao, F.; Brunelin, J.; Haffen, E. Can transcranial direct current stimulation (tDCS) alleviate symptoms and improve cognition in psychiatric disorders? *World J. Biol. Psychiatry* **2014**, *15*, 261–275. [CrossRef] [PubMed]

9.   Nitsche, M.A.; Fricke, K.; Henschke, U.; Schlitterlau, A.; Liebetanz, D.; Lang, N.; Henning, S.; Tergau, F.; Paulus, W. Pharmacological modulation of cortical excitability shifts induced by transcranial direct current stimulation in humans. *J. Physiol.* **2003**, *553*, 293–301. [CrossRef] [PubMed]

10.   Liebetanz, D.; Nitsche, M.A.; Tergau, F.; Paulus, W. Pharmacological approach to the mechanisms of transcranial DC-stimulation-induced after-effects of human motor cortex excitability. *Brain* **2002**, *125*, 2238–2247. [CrossRef] [PubMed]

11.   Das, S.; Holland, P.; Frens, M.A.; Donchin, O. Impact of transcranial direct current stimulation (tDCS) on neuronal functions. *Front. Neurosci.* **2016**, *10*, 550. [CrossRef] [PubMed]

12.   Kuo, M.-F.; Chen, P.-S.; Nitsche, M.A. The application of tDCS for the treatment of psychiatric diseases. *Int. Rev. Psychiatry* **2017**, *29*, 146–167. [CrossRef] [PubMed]

13.   Kuo, M.-F.; Paulus, W.; Nitsche, M.A. Therapeutic effects of non-invasive Brain Stimul.ation with direct currents (tDCS) in neuropsychiatric diseases. *Neuroimage* **2014**, *85 Pt 3*, 948–960. [CrossRef] [PubMed]

14.   Moher, D.; Liberati, A.; Tetzlaff, J.; Altman, D.G. PRISMA group preferred reporting items for systematic reviews and meta-analyses: The PRISMA statement. *J. Clin. Epidemiol.* **2009**, *62*, 1006–1012. [CrossRef] [PubMed]

15.   Ko, J.H.; Tang, C.C.; Eidelberg, D. Brain Stimul.ation and functional imaging with fMRI and PET. *Handb. Clin. Neurol.* **2013**, *116*, 77–95. [CrossRef] [PubMed]

16.   Nitsche, M.A.; Cohen, L.G.; Wassermann, E.M.; Priori, A.; Lang, N.; Antal, A.; Paulus, W.; Hummel, F.; Boggio, P.S.; Fregni, F.; et al. Transcranial direct current stimulation: State of the art 2008. *Brain Stimul.* **2008**, *1*, 206–223. [CrossRef] [PubMed]

17.   Nitsche, M.A.; Doemkes, S.; Karaköse, T.; Antal, A.; Liebetanz, D.; Lang, N.; Tergau, F.; Paulus, W. Shaping the effects of transcranial direct current stimulation of the human motor cortex. *J. Neurophysiol.* **2007**, *97*, 3109–3117. [CrossRef] [PubMed]

18.   Brunoni, A.R.; Moffa, A.H.; Sampaio-Júnior, B.; Gálvez, V.; Loo, C.K. Treatment-emergent mania/hypomania during antidepressant treatment with transcranial direct current stimulation (tDCS): A systematic review and meta-analysis. *Brain Stimul.* **2017**, *10*, 260–262. [CrossRef] [PubMed]

19.   Bikson, M.; Grossman, P.; Thomas, C.; Zannou, A.L.; Jiang, J.; Adnan, T.; Mourdoukoutas, A.P.; Kronberg, G.; Truong, D.; Boggio, P.; et al. Safety of transcranial direct current stimulation: Evidence based update 2016. *Brain Stimul.* **2016**, *9*, 641–661. [CrossRef] [PubMed]

20.   Cirillo, G.; Di Pino, G.; Capone, F.; Ranieri, F.; Florio, L.; Todisco, V.; Tedeschi, G.; Funke, K.; Di Lazzaro, V. Neurobiological after-effects of non-invasive Brain Stimul.ation. *Brain Stimul.* **2017**, *10*, 1–18. [CrossRef] [PubMed]

21. Brunoni, A.R.; Nitsche, M.A.; Bolognini, N.; Bikson, M.; Wagner, T.; Merabet, L.; Edwards, D.J.; Valero-Cabre, A.; Rotenberg, A.; Pascual-Leone, A.; et al. Clinical Research with Transcranial Direct Current Stimulation (Tdcs): Challenges and Future Directions. Available online: https://www.ncbi.nlm.nih.gov/pubmed/22037126 (accessed on 6 April 2018).

22. Dedoncker, J.; Brunoni, A.R.; Baeken, C.; Vanderhasselt, M.-A. A systematic review and meta-analysis of the effects of transcranial direct current stimulation (tDCS) over the dorsolateral prefrontal cortex in healthy and neuropsychiatric samples: Influence of stimulation parameters. *Brain Stimul.* **2016**, *9*, 501–517. [CrossRef] [PubMed]

23. Fregni, F.; Boggio, P.S.; Nitsche, M.A.; Marcolin, M.A.; Rigonatti, S.P.; Pascual-Leone, A. Treatment of major depression with transcranial direct current stimulation. *Bipolar Disord.* **2006**, *8*, 203–204. [CrossRef] [PubMed]

24. Boggio, P.S.; Rigonatti, S.P.; Ribeiro, R.B.; Myczkowski, M.L.; Nitsche, M.A.; Pascual-Leone, A.; Fregni, F. A randomized, double-blind clinical trial on the efficacy of cortical direct current stimulation for the treatment of major depression. *Int. J. Neuropsychopharmacol.* **2008**, *11*, 249–254. [CrossRef] [PubMed]

25. Boggio, P.S.; Bermpohl, F.; Vergara, A.O.; Muniz, A.L.C.R.; Nahas, F.H.; Leme, P.B.; Rigonatti, S.P.; Fregni, F. Go-no-go task performance improvement after anodal transcranial DC stimulation of the left dorsolateral prefrontal cortex in major depression. *J. Affect. Disord.* **2007**, *101*, 91–98. [CrossRef] [PubMed]

26. Loo, C.K.; Sachdev, P.; Martin, D.; Pigot, M.; Alonzo, A.; Malhi, G.S.; Lagopoulos, J.; Mitchell, P. A double-blind, sham-controlled trial of transcranial direct current stimulation for the treatment of depression. *Int. J. Neuropsychopharmacol.* **2010**, *13*, 61–69. [CrossRef] [PubMed]

27. Loo, C.K.; Alonzo, A.; Martin, D.; Mitchell, P.B.; Galvez, V.; Sachdev, P. Transcranial direct current stimulation for depression: 3-week, randomised, sham-controlled trial. *Br. J. Psychiatry* **2012**, *200*, 52–59. [CrossRef] [PubMed]

28. Rigonatti, S.P.; Boggio, P.S.; Myczkowski, M.L.; Otta, E.; Fiquer, J.T.; Ribeiro, R.B.; Nitsche, M.A.; Pascual-Leone, A.; Fregni, F. Transcranial direct stimulation and fluoxetine for the treatment of depression. *Eur. Psychiatry* **2008**, *23*, 74–76. [CrossRef] [PubMed]

29. Palm, U.; Schiller, C.; Fintescu, Z.; Obermeier, M.; Keeser, D.; Reisinger, E.; Pogarell, O.; Nitsche, M.A.; Möller, H.-J.; Padberg, F. Transcranial direct current stimulation in treatment resistant depression: A randomized double-blind, placebo-controlled study. *Brain Stimul.* **2012**, *5*, 242–251. [CrossRef] [PubMed]

30. Blumberger, D.M.; Tran, L.C.; Fitzgerald, P.B.; Hoy, K.E.; Daskalakis, Z.J. A randomized double-blind sham-controlled study of transcranial direct current stimulation for treatment-resistant major depression. *Front. Psychiatry* **2012**, *3*, 74. [CrossRef] [PubMed]

31. Bennabi, D.; Nicolier, M.; Monnin, J.; Tio, G.; Pazart, L.; Vandel, P.; Haffen, E. Pilot study of feasibility of the effect of treatment with tDCS in patients suffering from treatment-resistant depression treated with escitalopram. *Clin. Neurophysiol.* **2015**, *126*, 1185–1189. [CrossRef] [PubMed]

32. Valiengo, L.C.L.; Goulart, A.C.; de Oliveira, J.F.; Benseñor, I.M.; Lotufo, P.A.; Brunoni, A.R. Transcranial direct current stimulation for the treatment of post-stroke depression: Results from a randomised, sham-controlled, double-blinded trial. *J. Neurol. Neurosurg. Psychiatry* **2017**, *88*, 170–175. [CrossRef] [PubMed]

33. Brunoni, A.R.; Valiengo, L.; Baccaro, A.; Zanão, T.A.; de Oliveira, J.F.; Goulart, A.; Boggio, P.S.; Lotufo, P.A.; Benseñor, I.M.; Fregni, F. The sertraline vs. electrical current therapy for treating depression clinical study: Results from a factorial, randomized, controlled trial. *JAMA Psychiatry* **2013**, *70*, 383–391. [CrossRef] [PubMed]

34. Brunoni, A.R.; Sampaio-Junior, B.; Moffa, A.H.; Borrione, L.; Nogueira, B.S.; Aparício, L.V.M.; Veronezi, B.; Moreno, M.; Fernandes, R.A.; Tavares, D.; et al. The Escitalopram versus Electric Current Therapy for Treating Depression Clinical Study (ELECT-TDCS): Rationale and study design of a non-inferiority, triple-arm, placebo-controlled clinical trial. *Sao Paulo Med. J.* **2015**, *133*, 252–263. [CrossRef] [PubMed]

35. Loo, C.K.; Husain, M.M.; McDonald, W.M.; Aaronson, S.; O'Reardon, J.P.; Alonzo, A.; Weickert, C.S.; Martin, D.M.; McClintock, S.M.; Mohan, A.; et al. International Consortium of Research in tDCS (ICRT) International randomized-controlled trial of transcranial Direct Current Stimulation in depression. *Brain Stimul.* **2018**, *11*, 125–133. [CrossRef] [PubMed]

36. Martin, D.M.; Alonzo, A.; Ho, K.-A.; Player, M.; Mitchell, P.B.; Sachdev, P.; Loo, C.K. Continuation transcranial direct current stimulation for the prevention of relapse in major depression. *J. Affect. Disord.* **2013**, *144*, 274–278. [CrossRef] [PubMed]

37. Valiengo, L.; Benseñor, I.M.; Goulart, A.C.; de Oliveira, J.F.; Zanao, T.A.; Boggio, P.S.; Lotufo, P.A.; Fregni, F.; Brunoni, A.R. The sertraline versus electrical current therapy for treating depression clinical study (select-TDCS): Results of the crossover and follow-up phases. *Depress Anxiety* **2013**, *30*, 646–653. [CrossRef] [PubMed]

38. Kalu, U.G.; Sexton, C.E.; Loo, C.K.; Ebmeier, K.P. Transcranial direct current stimulation in the treatment of major depression: A meta-analysis. *Psychol. Med.* **2012**, *42*, 1791–1800. [CrossRef] [PubMed]

39. Berlim, M.T.; Van den Eynde, F.; Daskalakis, Z.J. Clinical utility of transcranial direct current stimulation (tDCS) for treating major depression: A systematic review and meta-analysis of randomized, double-blind and sham-controlled trials. *J. Psychiatr. Res.* **2013**, *47*, 1–7. [CrossRef] [PubMed]

40. Shiozawa, P.; Fregni, F.; Benseñor, I.M.; Lotufo, P.A.; Berlim, M.T.; Daskalakis, J.Z.; Cordeiro, Q.; Brunoni, A.R. Transcranial direct current stimulation for major depression: An updated systematic review and meta-analysis. *Int. J. Neuropsychopharmacol.* **2014**, *17*, 1443–1452. [CrossRef] [PubMed]

41. Meron, D.; Hedger, N.; Garner, M.; Baldwin, D.S. Transcranial direct current stimulation (tDCS) in the treatment of depression: Systematic review and meta-analysis of efficacy and tolerability. *Neurosci. Biobehav. Rev.* **2015**, *57*, 46–62. [CrossRef] [PubMed]

42. Brunoni, A.R.; Moffa, A.H.; Fregni, F.; Palm, U.; Padberg, F.; Blumberger, D.M.; Daskalakis, Z.J.; Bennabi, D.; Haffen, E.; Alonzo, A.; et al. Transcranial direct current stimulation for acute major depressive episodes: Meta-analysis of individual patient data. *Br. J. Psychiatry* **2016**, *208*, 522–531. [CrossRef] [PubMed]

43. Alonzo, A.; Chan, G.; Martin, D.; Mitchell, P.B.; Loo, C. Transcranial direct current stimulation (tDCS) for depression: Analysis of response using a three-factor structure of the Montgomery-Åsberg depression rating scale. *J. Affect. Disord.* **2013**, *150*, 91–95. [CrossRef] [PubMed]

44. D'Urso, G.; Dell'Osso, B.; Rossi, R.; Brunoni, A.R.; Bortolomasi, M.; Ferrucci, R.; Priori, A.; de Bartolomeis, A.; Altamura, A.C. Clinical predictors of acute response to transcranial direct current stimulation (tDCS) in major depression. *J. Affect. Disord.* **2017**, *219*, 25–30. [CrossRef] [PubMed]

45. Martin, D.M.; Yeung, K.; Loo, C.K. Pre-treatment letter fluency performance predicts antidepressant response to transcranial direct current stimulation. *J. Affect. Disord.* **2016**, *203*, 130–135. [CrossRef] [PubMed]

46. Brunoni, A.R.; Machado-Vieira, R.; Zarate, C.A.; Valiengo, L.; Vieira, E.L.; Benseñor, I.M.; Lotufo, P.A.; Gattaz, W.F.; Teixeira, A.L. Cytokines plasma levels during antidepressant treatment with sertraline and transcranial direct current stimulation (tDCS): Results from a factorial, randomized, controlled trial. *Psychopharmacology (Berl.)* **2014**, *231*, 1315–1323. [CrossRef] [PubMed]

47. Brunoni, A.R.; Machado-Vieira, R.; Zarate, C.A.; Vieira, E.L.M.; Valiengo, L.; Benseñor, I.M.; Lotufo, P.A.; Gattaz, W.F.; Teixeira, A.L. Assessment of non-BDNF neurotrophins and GDNF levels after depression treatment with sertraline and transcranial direct current stimulation in a factorial, randomized, sham-controlled trial (SELECT-TDCS): An exploratory analysis. *Prog. Neuropsychopharmacol. Biol. Psychiatry* **2015**, *56*, 91–96. [CrossRef] [PubMed]

48. Brunoni, A.R.; Machado-Vieira, R.; Zarate, C.A.; Vieira, E.L.M.; Vanderhasselt, M.-A.; Nitsche, M.A.; Valiengo, L.; Benseñor, I.M.; Lotufo, P.A.; Gattaz, W.F.; et al. BDNF plasma levels after antidepressant treatment with sertraline and transcranial direct current stimulation: Results from a factorial, randomized, sham-controlled trial. *Eur. Neuropsychopharmacol.* **2014**, *24*, 1144–1151. [CrossRef] [PubMed]

49. Brunoni, A.R.; Kemp, A.H.; Shiozawa, P.; Cordeiro, Q.; Valiengo, L.C.L.; Goulart, A.C.; Coprerski, B.; Lotufo, P.A.; Brunoni, D.; Perez, A.B.A.; et al. Impact of 5-HTTLPR and BDNF polymorphisms on response to sertraline versus transcranial direct current stimulation: Implications for the serotonergic system. *Eur. Neuropsychopharmacol.* **2013**, *23*, 1530–1540. [CrossRef] [PubMed]

50. Bennabi, D.; Pedron, S.; Haffen, E.; Monnin, J.; Peterschmitt, Y.; Van Waes, V. Transcranial direct current stimulation for memory enhancement: From clinical research to animal models. *Front. Syst. Neurosci.* **2014**, *8*, 159. [CrossRef] [PubMed]

51. Oliveira, J.F.; Zanão, T.A.; Valiengo, L.; Lotufo, P.A.; Benseñor, I.M.; Fregni, F.; Brunoni, A.R. Acute working memory improvement after tDCS in antidepressant-free patients with major depressive disorder. *Neurosci. Lett.* **2013**, *537*, 60–64. [CrossRef] [PubMed]

52. Moreno, M.L.; Vanderhasselt, M.-A.; Carvalho, A.F.; Moffa, A.H.; Lotufo, P.A.; Benseñor, I.M.; Brunoni, A.R. Effects of acute transcranial direct current stimulation in hot and cold working memory tasks in healthy and depressed subjects. *Neurosci. Lett.* **2015**, *591*, 126–131. [CrossRef] [PubMed]

53. Fregni, F.; Boggio, P.S.; Nitsche, M.A.; Rigonatti, S.P.; Pascual-Leone, A. Cognitive effects of repeated sessions of transcranial direct current stimulation in patients with depression. *Depress Anxiety* **2006**, *23*, 482–484. [CrossRef] [PubMed]

54. Wolkenstein, L.; Plewnia, C. Amelioration of cognitive control in depression by transcranial direct current stimulation. *Biol. Psychiatry* **2013**, *73*, 646–651. [CrossRef] [PubMed]

55. Salehinejad, M.A.; Ghanavai, E.; Rostami, R.; Nejati, V. Cognitive control dysfunction in emotion dysregulation and psychopathology of major depression (MD): Evidence from transcranial Brain Stimul.ation of the dorsolateral prefrontal cortex (DLPFC). *J. Affect. Disord.* **2017**, *210*, 241–248. [CrossRef] [PubMed]

56. Gögler, N.; Willacker, L.; Funk, J.; Strube, W.; Langgartner, S.; Napiórkowski, N.; Hasan, A.; Finke, K. Single-session transcranial direct current stimulation induces enduring enhancement of visual processing speed in patients with major depression. *Eur. Arch. Psychiatry Clin. Neurosci.* **2016**. [CrossRef] [PubMed]

57. Brennan, S.; McLoughlin, D.M.; O'Connell, R.; Bogue, J.; O'Connor, S.; McHugh, C.; Glennon, M. Anodal transcranial direct current stimulation of the left dorsolateral prefrontal cortex enhances emotion recognition in depressed patients and controls. *J. Clin. Exp. Neuropsychol.* **2017**, *39*, 384–395. [CrossRef] [PubMed]

58. Bueno, V.F.; Brunoni, A.R.; Boggio, P.S.; Bensenor, I.M.; Fregni, F. Mood and cognitive effects of transcranial direct current stimulation in post-stroke depression. *Neurocase* **2011**, *17*, 318–322. [CrossRef] [PubMed]

59. Brunoni, A.R.; Tortella, G.; Benseñor, I.M.; Lotufo, P.A.; Carvalho, A.F.; Fregni, F. Cognitive effects of transcranial direct current stimulation in depression: Results from the SELECT-TDCS trial and insights for further clinical trials. *J. Affect. Disord.* **2016**, *202*, 46–52. [CrossRef] [PubMed]

60. Tortella, G.; Casati, R.; Aparicio, L.V.M.; Mantovani, A.; Senço, N.; D'Urso, G.; Brunelin, J.; Guarienti, F.; Selingardi, P.M.L.; Muszkat, D.; et al. Transcranial direct current stimulation in psychiatric disorders. *World J. Psychiatry* **2015**, *5*, 88–102. [CrossRef] [PubMed]

61. Chew, T.; Ho, K.-A.; Loo, C.K. Inter- and intra-individual variability in response to transcranial direct current stimulation (tDCS) at varying current intensities. *Brain Stimul.* **2015**, *8*, 1130–1137. [CrossRef] [PubMed]

62. Martin, D.M.; Alonzo, A.; Mitchell, P.B.; Sachdev, P.; Gálvez, V.; Loo, C.K. Fronto-extracephalic transcranial direct current stimulation as a treatment for major depression: An open-label pilot study. *J. Affect. Disord.* **2011**, *134*, 459–463. [CrossRef] [PubMed]

63. Bai, S.; Dokos, S.; Ho, K.-A.; Loo, C. A computational modelling study of transcranial direct current stimulation montages used in depression. *Neuroimage* **2014**, *87*, 332–344. [CrossRef] [PubMed]

64. Ho, K.-A.; Bai, S.; Martin, D.; Alonzo, A.; Dokos, S.; Loo, C.K. Clinical pilot study and computational modeling of bitemporal transcranial direct current stimulation, and safety of repeated courses of treatment, in major depression. *J. ECT* **2015**, *31*, 226–233. [CrossRef] [PubMed]

65. Ho, K.-A.; Bai, S.; Martin, D.; Alonzo, A.; Dokos, S.; Puras, P.; Loo, C.K. A pilot study of alternative transcranial direct current stimulation electrode montages for the treatment of major depression. *J. Affect. Disord.* **2014**, *167*, 251–258. [CrossRef] [PubMed]

66. Pavlova, E.L.; Menshikova, A.A.; Semenov, R.V.; Bocharnikova, E.N.; Gotovtseva, G.N.; Druzhkova, T.A.; Gersamia, A.G.; Gudkova, A.A.; Guekht, A.B. Transcranial direct current stimulation of 20- and 30-minutes combined with sertraline for the treatment of depression. *Prog. Neuropsychopharmacol. Biol. Psychiatry* **2018**, *82*, 31–38. [CrossRef] [PubMed]

67. Fricke, K.; Seeber, A.A.; Thirugnanasambandam, N.; Paulus, W.; Nitsche, M.A.; Rothwell, J.C. Time course of the induction of homeostatic plasticity generated by repeated transcranial direct current stimulation of the human motor cortex. *J. Neurophysiol.* **2011**, *105*, 1141–1149. [CrossRef] [PubMed]

68. Alonzo, A.; Brassil, J.; Taylor, J.L.; Martin, D.; Loo, C.K. Daily transcranial direct current stimulation (tDCS) leads to greater increases in cortical excitability than second daily transcranial direct current stimulation. *Brain Stimul.* **2012**, *5*, 208–213. [CrossRef] [PubMed]

69. Ferrucci, R.; Bortolomasi, M.; Vergari, M.; Tadini, L.; Salvoro, B.; Giacopuzzi, M.; Barbieri, S.; Priori, A. Transcranial direct current stimulation in severe, drug-resistant major depression. *J. Affect. Disord.* **2009**, *118*, 215–219. [CrossRef] [PubMed]

70. Brunoni, A.R.; Ferrucci, R.; Bortolomasi, M.; Vergari, M.; Tadini, L.; Boggio, P.S.; Giacopuzzi, M.; Barbieri, S.; Priori, A. Transcranial direct current stimulation (tDCS) in unipolar vs. bipolar depressive disorder. *Prog. Neuropsychopharmacol. Biol. Psychiatry* **2011**, *35*, 96–101. [CrossRef] [PubMed]

71. Dondé, C.; Amad, A.; Nieto, I.; Brunoni, A.R.; Neufeld, N.H.; Bellivier, F.; Poulet, E.; Geoffroy, P.-A. Transcranial direct-current stimulation (tDCS) for bipolar depression: A systematic review and meta-analysis. *Prog. Neuropsychopharmacol. Biol. Psychiatry* **2017**, *78*, 123–131. [CrossRef] [PubMed]

72. Minhas, P.; Bikson, M.; Woods, A.J.; Rosen, A.R.; Kessler, S.K. Transcranial direct current stimulation in pediatric brain: A computational modeling study. *Conf. Proc. IEEE Eng. Med. Biol. Soc.* **2012**, *2012*, 859–862. [CrossRef] [PubMed]

73. Moliadze, V.; Schmanke, T.; Andreas, S.; Lyzhko, E.; Freitag, C.M.; Siniatchkin, M. Stimulation intensities of transcranial direct current stimulation have to be adjusted in children and adolescents. *Clin. Neurophysiol.* **2015**, *126*, 1392–1399. [CrossRef] [PubMed]

74. Kessler, S.K.; Minhas, P.; Woods, A.J.; Rosen, A.; Gorman, C.; Bikson, M. Dosage considerations for transcranial direct current stimulation in children: A computational modeling study. *PLoS ONE* **2013**, *8*, e76112. [CrossRef] [PubMed]

75. Brunoni, A.R.; Ferrucci, R.; Bortolomasi, M.; Scelzo, E.; Boggio, P.S.; Fregni, F.; Dell'Osso, B.; Giacopuzzi, M.; Altamura, A.C.; Priori, A. Interactions between transcranial direct current stimulation (tDCS) and pharmacological interventions in the Major Depressive Episode: Findings from a naturalistic study. *Eur. Psychiatry* **2013**, *28*, 356–361. [CrossRef] [PubMed]

76. Segrave, R.A.; Arnold, S.; Hoy, K.; Fitzgerald, P.B. Concurrent cognitive control training augments the antidepressant efficacy of tDCS: A pilot study. *Brain Stimul.* **2014**, *7*, 325–331. [CrossRef] [PubMed]

77. Brunoni, A.R.; Boggio, P.S.; De Raedt, R.; Benseñor, I.M.; Lotufo, P.A.; Namur, V.; Valiengo, L.C.L.; Vanderhasselt, M.A. Cognitive control therapy and transcranial direct current stimulation for depression: A randomized, double-blinded, controlled trial. *J. Affect. Disord.* **2014**, *162*, 43–49. [CrossRef] [PubMed]

78. Martin, D.M.; Teng, J.Z.; Lo, T.Y.; Alonzo, A.; Goh, T.; Iacoviello, B.M.; Hoch, M.M.; Loo, C.K. Clinical pilot study of transcranial direct current stimulation combined with Cognitive Emotional Training for medication resistant depression. *J. Affect. Disord.* **2018**, *232*, 89–95. [CrossRef] [PubMed]

79. Mayur, P.; Howari, R.; Byth, K.; Vannitamby, R. Concomitant transcranial direct current stimulation with ultrabrief electroconvulsive therapy: A 2-week double-blind randomized sham-controlled trial. *J. ECT* **2018**. [CrossRef] [PubMed]

MDPI

Article

From e-Health to i-Health: Prospective Reflexions on the Use of Intelligent Systems in Mental Health Care

Xavier Briffault [1], Margot Morgiève [2,*] and Philippe Courtet [2,3,4]

[1] Centre de Recherche Médecine, Sciences, Santé, Santé Mentale, Société (CERMES3), UMR CNRS 8211-Unité Inserm 988-EHESS-Université Paris Descartes, 75006 Paris, France; xavier.briffault@parisdescartes.fr

[2] FondaMental Foundation, 94000 Créteil, France; philippe.courtet@univ-montp1.fr

[3] Institut National de la Santé et de la Recherche Médicale U1061 Neuropsychiatry: Epidemiological and Clinical Research, University of Montpellier, 34000 Montpellier, France

[4] Department of Psychiatric Emergency & Acute Care, Lapeyronie Hospital, CHU Montpellier, 34000 Montpellier, France

* Correspondence: margotmorgieve@yahoo.fr; Tel.: +33-6-0959-6532

Received: 25 April 2018; Accepted: 28 May 2018; Published: 31 May 2018

Abstract: Depressive disorders cover a set of disabling problems, often chronic or recurrent. They are characterized by a high level of psychiatric and somatic comorbidities and represent an important public health problem. To date, therapeutic solutions remain unsatisfactory. For some researchers, this is a sign of decisive paradigmatic failure due to the way in which disorders are conceptualized. They hypothesize that the symptoms of a categorical disorder, or of different comorbid disorders, can be interwoven in chains of interdependencies on different elements, of which it would be possible to act independently and synergistically to influence the functioning of the symptom system, rather than limiting oneself to targeting a hypothetical single underlying cause. New connected technologies make it possible to invent new observation and intervention tools allowing better phenotypic characterization of disorders and their evolution, that fit particularly well into this new "symptoms network" paradigm. Synergies are possible and desirable between these technological and epistemological innovations and can possibly help to solve some of the difficult problems people with mental disorders face in their everyday life, as we will show through a fictional case study exploring the possibilities of connected technologies in mental disorders in the near future.

Keywords: m-health; i-health; depression; nosography; categorizations; symptoms networks; ecological momentary assessment; ecological momentary intervention; fictional case study

1. Introduction

1.1. Context

The notion of "depressive disorders" covers a set of disabling problems, often chronic or recurrent, whose prevalence in the general population is high. They are generally characterized by a high level of psychiatric and somatic comorbidities and a high proportion of chronicity or relapses. For these reasons, they represent an important public health problem and considerable suffering for the people who endure them and their families. To date, therapeutic solutions remain unsatisfactory, despite the intensity of research on this subject for decades. Many researchers today consider that this lack of satisfactory results is only temporary and that decisive therapeutic innovations may emerge in the future within the current paradigm, despite the fact that such results have not been obtained in several decades. Others, in contrast, see it as a sign of a decisive problem in the paradigm in which disorders are conceptualized. One of the central problems is that, although the particularly multifactorial, "bio-psycho-social" nature of these disorders—like many other mental disorders—is the subject of international consensus, it remains classic, however, to consider these multiple factors only as contributing to the development of an underlying cause of an isolable disease, the symptoms of which would only be the observable effect of that cause. The various symptoms of depression would thus be the effect of "depression", just as the multiple symptoms of syphilis are the effect of pale treponemal infection, which is the sole initial cause.

This organizational scheme is particularly visible in the Diagnostic and Statistical Manual of Mental Disorders (DSM) [1]. This reference nosography of the American Psychiatric association (APA), which has been organizing international research on mental disorders for several decades, adopted in its third version (DSM-III, published in the early 1980s) a categorical definition of disorders independent of their context of occurrence. Disorders are constructed by composition of elementary observable symptoms not articulated with each other or with context to obtain a polythetic definition (i.e., composed of a "mandatory" core of symptoms to which are added "optional" symptoms). This syndrome is supposed to represent the theoretical level immediately above the elementary symptoms, each definition being the supposedly reliable formalization (i.e., of a high degree of intersubjective agreement) of phenomenal random variations statistically observed around a supposed natural morbid entity which would be the valid denotation [2,3].

This approach to mental disorders is more generally part of a "medical model" of disorders [4]. In this medical model the patient is thought of under the angle of a specific problem that he has, and which can be circumscribed and isolated, this problem being directly caused by one (at most some) underlying cause(s), these causes directly explaining the problem. This conceptualization directs research towards the hypothesis that there are therapeutic ingredients specific to each of these causes. Each of these ingredients having a specific efficacy that must be evaluated by randomized controlled trials specifically constructed to evaluate the intrinsic efficacy of an intervention independently of any context. For several decades, this approach has produced an important list of empirically supported treatments (i.e., for which a statistically significant positive effect size has been calculated in randomized controlled trials), whether pharmacological, psychotherapeutic, neurosurgical, and so on, treatments described as precisely as possible in protocol manuals defined independently of the context. While it is undeniable that these multiple treatments have therapeutic efficacy, the fact remains that this efficacy is still low and that many disorders and patients are not or only slightly improved by these approaches [5,6].

The normative ideal of an isolable mental disorder with a single cause massively organizes the collective functioning of research and practice on mental disorders. It prevents significant interactions between symptoms from being taken into consideration and syndromic entities used to characterize mental disorders from being considered as something other than the manifestations of the natural entity that is supposed to cause them. It is also massively structuring possible therapeutic practices, because it prohibits the hypothesis that the symptoms of a categorical disorder, or of different comorbid disorders, can be interwoven in chains of interdependencies. It thus prevents to conceive intervention acting independently and synergistically on the elements of theses chains to influence the functioning of the symptom system, rather than limiting oneself to targeting a hypothetical single underlying cause.

This new paradigm, thought still in its infancy and under debate, appears as an interesting framework for conceptualizing mental disorders, particularly depressive disorders. If we adopt this paradigm, there is an important need to rethink the conceptual bases of physio-psycho-socio-pathological knowledge that underlie the conceptions of possible therapeutic approaches, and to invent new observation tools allowing a better phenotypic characterization of disorders and their evolution.

1.2. Objectives

The objective of this article is to propose some elements for reflection on possible synergies between recent developments in the conceptualization of mental disorders, particularly depressive disorders and innovations that are developing very fast in the field of connected objects and applications commonly grouped under the terms e-health/m-health. More precisely, we propose a systematised yet speculative use case of these new technologies based on a fictional case study of a person presenting complex mental health problems. Our main goal with this case is to propose a concrete example of how intelligent connected information systems could be useful in clinical and therapeutical settings in order to ecologically augment clinical observations and therapeutical interventions in mental health. Moreover, while empirical analysis of the current uses of Ecological Momentary Assessment (EMA)/ Ecological Momentary Intervention (EMI) devices is essential, the conception of these devices is still limited by conceptual limitations that prevents such empirical analyses to serve as a basis for an in-depth improvement of such tools. Consequently, it is also necessary to anticipate the upheavals that will occur in the near future by reflecting on possible, but not yet proven, uses of these devices for people with psychological problems/mental disorders, long before it can, eventually if proven safe and useful, be proposed in clinical routines.

2. New Approaches in Conceptualizing Mental Disorders

A current of research whose foundations can be traced back to the end of the 2000s [7,8] and which has largely developed since then [3,9] is fully in line with the paradigmatic rupture that we evoke. It proposes a new conceptual and therapeutic approach that makes it possible to unessentialize mental disorders by redefining them as clusters of properties connected between them by a homeostasic system of causal relations, thus going over the limits of the DSM that we have just mentioned. The disorders are then conceived as stable attractors in a network of properties, which emerge from the dynamic organization of causal links within the network rather than being arbitrarily reified into independent static entities. This type of networked approach allows for the adoption of "a psychosystemic approach that addresses the inherent complexity of mental disorders by using explicit models of the interactions between their psychological, biological and social characteristics that play a role in the development of psychiatric conditions, understood as clusters of causally related properties" [10]. As a result, the first category of apprehension of persons with mental disorders is no longer limited to the isolated, monadic individual suffering from an isolated disease caused by a single cause. Instead, today it tends to evolve towards an agent in situation whose action is disturbed by inadequacies between his characteristics and those of his situation, characterizable in real time by multiple parameters in mutual interactions articulated to the parameters of the situation.

In addition to the "internal" causal relationships linking the symptoms of a given psychiatric disorder together, these new approaches also make it possible to integrate the dynamic couplings between the modes of functioning internal to the individual and those that organize social, relational, physical environments and situations … of which he is part, with a specific interest in the "depressogenic", "anxiogenic", "OCDgenic", "autistogenic", "suicidogenic", "disadaptogenic", "burn-outogenic" … properties of these couplings. The pathogenic or disabling dimension here no longer belongs solely to the individual or solely to the situation but to the joints of both. This logic no longer looks only at the causes, but also at the consequences in a given environment of health problems in order to classify them [11]. In this shift from the patient with a psychiatric-neurological illness to the "person in situation", the medical objective of reducing or eliminating symptoms loses its centrality. It becomes one of the elements of an approach aimed at reducing the daily impact of these symptoms in order to improve the quality of life [12–14]. A shift is thus taking place from a strictly curative logic inscribed in a medical model to a rehabilitation logic also inscribed in a contextual model. The therapeutic approach aims at the reduction of functional consequences only by the reduction/suppression of the pathology. The new situated approaches additionally open the possibility of acting directly on the disabling consequences in ecological situation, not only by looking at the "big" effects of the "big" pathological entities, but also and above all by looking at the articulations of individual micro-mechanisms and socio-environmental micro-mechanisms in order to be able to intervene finely on them.

These conceptual developments in mental health are thus no longer based solely on large diagnostic entities (depression, schizophrenia, etc.) but on pluralities of local dysfunctions or problematic stabilizations in graphs of interactions of "symptoms" inscribed in a process [15–17]. The focus is no longer limited to the single "person" pole but extends to its functioning in a real environment [18]. They make it possible to imagine new forms of action that are finer, more contextualized, more ecological, more interactive, more focused on changes in interactions between the various components of the person and the various components of his or her real environment.

3. New Opportunities for Observation

These innovations in the conceptualization of mental disorders open up new clinical and therapeutic possibilities. It is now theoretically possible not only to observe with much finer granularity the components of a mental disorder and their internal interactions with the components of the situation and the environment, but also to intervene finely on these components and on the causal chains that maintain the system in patterns that are pejorative for the person. To these new theoretical possibilities, different emerging technologies offer operational implementation possibilities.

This is particularly the case for connected mobile technologies that are developing in the e-health/m-health field. Connected technologies, particularly because of their very small size and ubiquitous communication capabilities, offer hitherto completely unprecedented possibilities for collecting multiple data in real time [19–21] associated with a space-time location on a person and her proximal or distal environment [22], but also to intervene at multiple levels on this person and this environment by coupling in real time the interventions to the collected data. To date, no other therapy, care or support system has had such an ability to integrate into the subject's life in order to gather information and intervene, whether it is the classic office consultation, home visits, telemedicine, or even complete hospitalization, which certainly allows total observation of the patient, but has nothing ecological about it. By making possible the simultaneous use, in real time and in ecological situations, of multiple micro-sensors, micro-effectors and human actors, all interconnected, the new connected technological devices offer unprecedented possibilities for reducing the grain of data collection and interventions on the person. The fineness of the possible articulations of individual parameters with each other and with multiple situational parameters, as well as the spatial and temporal perimeter of the accessible information [23], is thereby massively increased.

4. Synergies between Conceptual and Technological Innovations

This is why the epistemological evolutions we have just mentioned-"network" approaches to mental disorders-and the evolutions of connected technologies in the field of health (e-health, m-health)-objects connected in networks-tend today (although obviously coming from different epistemic universes) to join. It allow in mental health problems micro-interventions much more targeted, evolutive, and contextualized than allowed by the previous approaches applying "macro-interventions" to "macro-pathologies" [24]. The patient is no longer seen as a monadic individual altered by diseases, occasionally encountered out of context by practitioners who have only clinical and paraclinical observations, supplemented by the patient's own clinical and paraclinical observations, accessible in the limited space of the consultation for any information on the patient's functioning. Instead, the patient tends today to evolve towards an agent in a situation whose action is disturbed by inadequacies between his functions and those of his situation, characterizable in real time by multiple parameters in mutual interactions articulated to the parameters of the situation [16,17,25], on which multiple people using multiple technical devices (applications and connected objects) can now intervene in real time and in an ecological environment.

We now have the ability to generate in real time a "cloud of data" associated with the person and his situation, to statistically analyze co-variations in order to produce inferences based on "data-based" or "data-driven" predictive models. We are also able to integrate these data directly into a priori models (when available) of the relationships between the measured variables, in order to produce "model-based" predictions, and to make all these elements available to the patient and his therapists, in the service of an "augmented" clinic and therapy [26]. Among these multiple parameters whose measurement is now possible, some can be considered as items on which we wish to act specifically, either directly, or through causal factors, or assumed to be causal on the basis of strong statistical correlations derived from analysis of groups or intra-individual correlations. In the case of mental disorders, one can for example wish to act on mood, anxiety, fear, stress, sleep, arousal, motivation, sedentary life, mental suffering, aversive tensions, suicidal ideations, obsessions, compulsions, consumption of psychotropic substances, eating habits, social relations, hallucinations, executive functions, the presence to oneself and to the environment, muscular tensions, memory, reasoning, somatoform symptoms [27–32]. In short, on any psychological, physiological, behavioural parameter . . . considered problematic and within the scope of "mental health", and for which there would be a direct means of action or an upstream influencing factor. A considerable number of applications have been developed in the field of e-mental health, to the point that it can almost be said that whatever the problem is considered "there is an application for that", according to the title of a recent article [33]. This is particularly the case for depressive disorders [29,34]. While most of these applications are not evaluated, available results show that these technologies can be not only effective but also cost-effective [35–37].

5. From e-Mental Health to i-Mental Health

However, these applications generally continue to fall within the "isolable disorder/specific intervention" paradigm, the limits of which have been shown and do not allow the best use to be made of isomorphism between a network approach to disorders and a network approach to connected objects. There is therefore reason to fear that the phenomenon of conceptual limitation observed in the design of past therapies may be repeated in the future for these new technologies if conceptual frameworks do not evolve sufficiently. This is the hypothesis put forward by various authors, including Berrouiguet et al. [38], who call for the "e-health" paradigm to be overcome by a broader approach they call "i-health" (intelligent health), or Briffault et al. who introduce the concept of the Technologically Augmented Clinical and Therapeutic relationship (TACT) [24,39,40].

According to these authors, the usual approach to psychiatric care is limited in its possibilities of data collection, diagnosis and intervention by the very characteristics of consultations between practitioners and patients, brief, rare, taking place outside any real life context and in which relatives rarely participate. The tools currently proposed in the field of e-health certainly make it possible to extend the collection of data during the periods between consultations, by Internet or by mobile phone and to store them in personal electronic files. But this collection remains however more often than not punctual, it requires an intervention of the patient, does not allow the collection in real time of personal and contextual information. In truth, it does not qualitatively improve the possibilities of medical analysis.

Now, what connected objects and the applications that accompany them offer today is the possibility of interconnecting in real time the multiple components of the person in action. These technologies offer the possibility of articulating: vertically the macro (social, smart cities . . .), meso (situation, smart homes, physical and relational environment) and micro (the person) and even very micro (different physiological and psychological parameters and mechanisms) levels; and horizontally the different domains of existence (nutrition, activities, relationships, movements, organization, cognitions, affects . . .) involved in daily living. All elements whose proper functioning and articulation are disturbed by mental disorders and can contribute to their persistence or reduction.

These new devices thus offer new possibilities for spatial, temporal, and thematic extensions of psychiatric, psychotherapeutic, psycho-educational, medical relationships . . . , whose observation and intervention potential can now be extended far beyond consultation to potentially concern all areas of life, at any time and in any place, while integrating the possibilities for automatic analysis and analysis assistance offered by data-mining and deep-learning technologies. As Berrouiguet et al. point out, this is the beginning of a twofold evolution [38]. First of all, the routine integration of the data collected in the psychiatric consultation situation will give practitioners real-time access to subjective and behavioural data from patients and their families, which will significantly modify their clinical observation and therapeutic intervention possibilities. Then, the addition to the "handmade" analysis of the data by the practitioner-made impossible by the amount of data generated-of automatic analysis possibilities will allow the development of intelligent medical decision support systems allowing the development of predictive models and personalized treatments.

This is no longer only a quantitative evolution, but a real qualitative revolution. From the cylinder-rolled sheets of Laennec's first stethoscope to the most sophisticated MRIs or biological analyses, and even current Ecological Momentary Assessment applications, there is no real qualitative leap. Tools of "mediate auscultation", these technologies only improve its possibilities, without modifying its basis: it is always the physician who makes sense of the observed data, in the clinical context integrated into his relationship with the patient. Never has a scanner, a sheet of biological results, a psychiatric evaluation scale or its smart phone version attempted to analyze by themselves what they were giving the doctor as data on the patient's condition. But this is what connected technologies combined with artificial intelligence technologies allow today, which justifies their being considered "disruptive". Indeed, they lead to a break in continuity in the status of observation and intervention tools in medicine and mental medicine; they move them from the stage of tools for extending the possibilities of the doctor to the stage of tools capable of proposing interpretations or even making decisions and carrying out actions themselves, in interaction with the various stakeholders in a care process (people with disorders, mental health practitioners, relatives, etc.).

6. What Elements for an i-Mental-Health System

To provide the best possible service to people with mental health disorders, the i-mental-health devices of the future will need to build on both the technological and conceptual innovations we have presented. They should take note of the fact that the health parameters involved in the functioning of a person in a situation are multiple and diverse, that they are in complex and dynamic interactions, that these interactions concern parameters belonging to different levels (genetic biological, psychological, situational, contextual, social) that influence each other inter and intra-level, and that the system disturbances that constitute the relationships between these health parameters have functional consequences that depend on the individual in the situation. The value attributed to them is also individual and may vary according to times and situations within the same individual. The network of interactions between health parameters can evolve very gradually or very suddenly towards states with pejorative functional consequences, or even towards morbid states that stabilize and can be considered as characterized disorders [17,26,41].

An i-mental health system must offer its users (professionals, patients, relatives) a set of health parameters, enable them to select those they wish to monitor according to their characteristics and their situation. It must also provide them with the means to highlight the fine relationships between these health parameters in order to be able to determine relevant places and means of intervention, based as much on the possibilities of automatic data-based and model-based analyses as on the possibilities of a clinical augmentation that these new tools/conceptualizations offer to users.

7. What Observation Tools

Any health parameter selected as a relevant observable must be capable of being monitored by an adapted observation tool. In order to increase compliance and to preserve the ecological validity of the data collection, the use of tools must be as simple and transparent as possible and be able to fit into the flow of activity without disrupting it. For example, in all possible cases, objective passive sensors that do not require any user action should be preferred to the use of questionnaires that consume time and disrupt the flow of activity, and that are also subject to recall bias and various cognitive and desirability biases.

Nevertheless, the use of questionnaires remains unavoidable in certain cases, typically for parameters related to user subjectivity. For example, no passive sensor can measure the degree of psychological suffering, assess quality of life, evaluate the level of aversive tension or the feeling of having a meaningful life, and so on. At the other end of the spectrum, no subjective assessment can better than an objective measure account for the degree of activity (actimeter), heart rate (heart rate monitor), blood pressure (blood pressure monitor), sleep characteristics (actimeter + heart rate monitor + other specific sensors of brain activity), biological markers, and so on.

However, the subjective consequences of variations in an objective parameter remain within the user's appreciation and expression. As a simple example, a tennis player and a footballer will not give the same value to a wrist tendonitis, even if the pathology is "the same". Some parameters also have intermediate statuses in that they can be observed by different complementary means. An example is depression. If it can be evaluated by self-administered or hetero-administered standardized questionnaires (Hamilton scale [42], Beck scale [43], PHQ-9 [44] . . .), simplified subjective assessments (visual analogue scales of depressive feelings [45]), it is also clinically - and therefore more or less objectively - observable, whether it is by clinical collection by a trained professional, or proxies whose predictive value is more or less good (different parameters of voice, activity, motor skills, relationships, lexical and semantic content of communications, etc.) [46–49]. A conjunction of observation instruments therefore makes it possible to triangulate the observed value for this type of parameter.

8. Process for Designing and Using an i-Mental-Health Device

The development of i-mental health systems integrating the functionalities that have been presented is a complex process, which is beginning to be the subject of new work [50–52]. An integrated device will necessarily have to rely on a team of health professionals trained in the use of connected devices, able to propose a set of functionalities adapted to each user according to his needs and situation, to analyse the data collected and to propose interventions. These professionals will need to be able to rely on a platform that allows them to generate an integrated suite of mobile applications in a totally flexible way to the patient/therapist's choice by integrating pre-existing or custom-developed functionalities that can be reprogrammed according to the patient's evolution. This suite should be capable of: (a) collecting health data in programmable ways; (b) using data from diverse eHealth/wellness applications; (c) aggregating standardized sensor flows and commercial applications; (d) providing therapeutic and prevention support and accompaniment tools; (e) dialogue with environmental servers (smart cities, smart homes, etc.); and (f) proposing user interfaces allowing data sharing for co-analysis and co-decision purposes between users, healthcare professionals, relatives, other stakeholders, and so on.

Co-decision on the choice of observation and co-design tools for the intervention programme involves the patient and a multidisciplinary team of health professionals, specially trained in the use of connected health tools in a consultation (which can be carried out remotely). In collaboration with the patient, this involves formalizing the main relevant elements of his health problems and situation in order to set up a complete observation and intervention application based on health parameters, questionnaires, connected objects and interventions available within the device. In order for the consultation to be as effective as possible, rapid training in the principles of prevention/promotion of health that will be used can be offered prior to the consultation to the patient in the form of an e-training/psychoeducation tool. If there are specific needs not covered by existing modules, specific developments may be implemented. They are then genericized and integrated into the basic proposal. The patient should ideally be able to connect to the software device specifically generated for him on his various computer hardware (computer, tablet, smartphone, connected objects) as soon as the consultation is over, and follow a quick (tele)training course on its use. The patient then lives his life using the different applications, connected objects, questionnaires that are proposed to him, analyses his data using the graphic tools made available to him, and implements the actions that are proposed to him by his device. He may contact the multidisciplinary team at his own initiative, or on the proposal of the system, to examine specific health problems highlighted by the use of the device, which may lead to changes in the device (new observations, new interventions). At the same time, the health team can contact the patient, based on a human and/or automated analysis of his data, in order to propose new interventions.

The data collected by the observation devices and the interventions implemented are stored in the database, thus allowing individual real-time, localized longitudinal monitoring. Data aggregation for all users provides a representation of user health parameters and behaviors and provides objective information for inter-individual comparisons and quasi-experimental evaluations.

9. A Fictional Case Study

The (mental) e-health field is particularly active, producing new devices at an extremely rapid pace. While empirical analysis of the current uses of these devices is essential, it is also necessary to anticipate the upheavals that will occur in the near future by reflecting on possible, but not yet proven, uses of these devices for people with psychological problems/mental disorders. Fictive case analysis offers such opportunities to reflect in advance on emerging technologies whose "disruptive" potential requires that reflections are not limited to known uses. This is what we propose here with a constructed case of depressive problem in which we imagine the possible design and use processes of a device for a Technologically Augmented Clinical and Therapeutic relationship (TACT), (Figures 1 and 2).

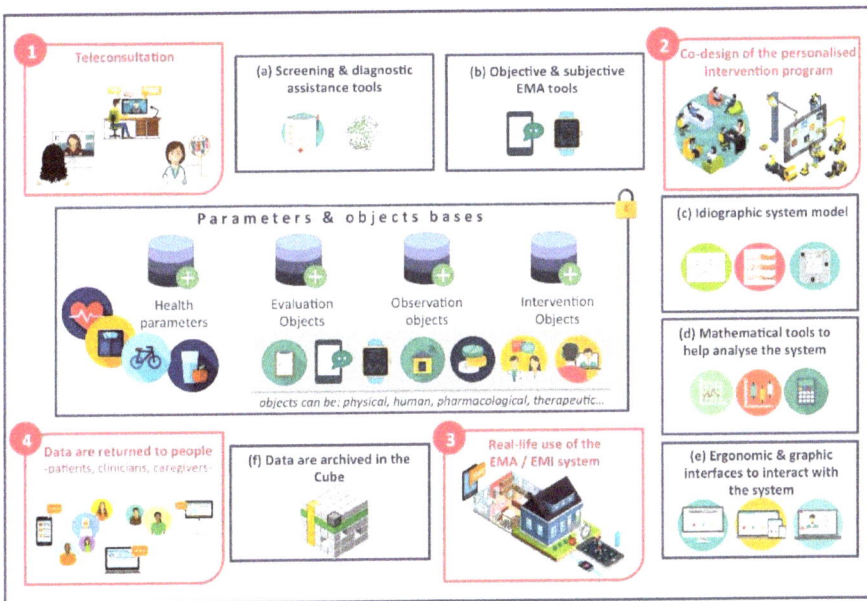

Figure 1. Design and use processes of a device for a Technologically Augmented Clinical and Therapeutic relationship.

Bob is a married man in his fifties, with two children, working on a stable job in an insurance company, living in a remote rural area, 90 min by car from the nearest metropolis, with few medical resources and even fewer psychological and psychiatric resources, a situation in which the use of connected devices is particularly suitable [53]. Confronted for several months with recurrent back and abdominal pain, palpitations, sleep problems and fits of profuse sweats, he consulted his general practitioner. Neither the clinical examination nor the paraclinical investigations carried out make it possible to highlight somatic causes to the symptoms of which Bob complains. His doctor then suggests that he consults a multi-professional e-mental health team to which he can give access, with his consent, to the data he has already collected. The financial coverage of the care protocol that will be put in place will be provided by Bob's supplementary insurance, funded by his company as part of a collective psychosocial risk prevention program.

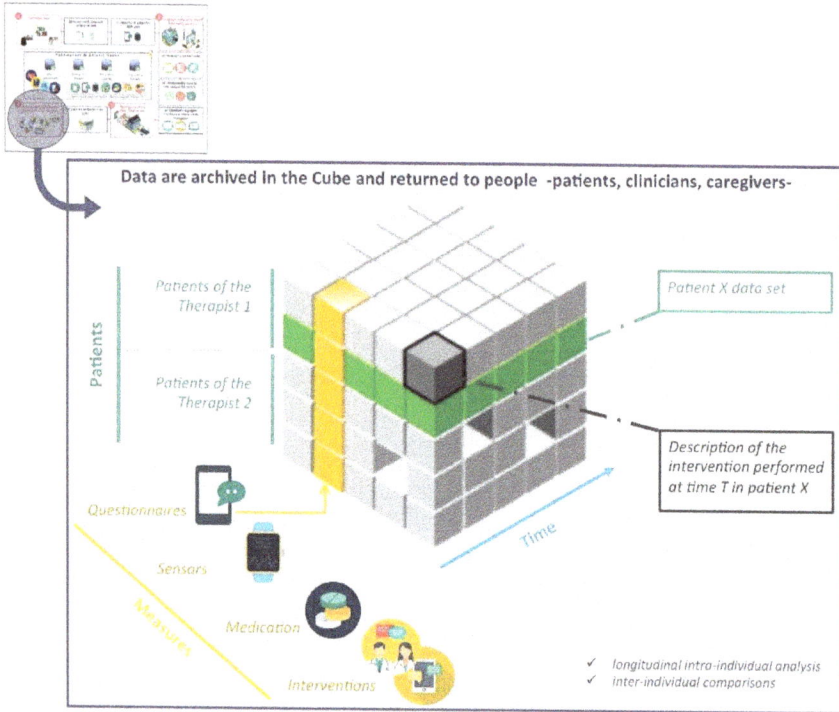

Figure 2. Patients X measures X time database.

The data already collected and an initial teleconsultation interview (Figure 1, step 1) with Bob guide the team towards an initial set of hypotheses on causes related to the anxiety-depressive constellation of symptoms Bob complains about. In addition to the symptoms in the foreground that motivated the consultation with the general practitioner, the interview shows that Bob feels depressed. He cannot do even the smallest project, he does not want anything, he cannot have pleasure in anything. He has relationship difficulties that are unusual for him, can no longer get in touch with others, and feels that people are moving away from him. He feels in his work unbearable psychological pressures, feels controlled and monitored constantly, has the feeling of growing dehumanization, of being treated as a number and having to treat his clients as such. Although he thinks that it is his company that has changed and not his point of view, he also says he prefers to abandon all responsibility to content himself with subordinate tasks rather than continue to occupy a position that is nevertheless more rewarding and meaningful, but that requires the mobilization of cognitive, emotional and executive resources that he no longer seems to have. The interview also shows that Bob attributes his difficulties not only to the objective evolution of his work context, but also partly to two things that do not depend on it. On the one hand a painful event, the premature death of his mother following an illness when he was a young teenager. He considers that he has not mourned this very protective mother who, in his opinion, has not made him mentally strong enough and equipped him with the skills to face life's difficulties on his own. Her advice, her presence, her support are missing more and more painfully with each problem encountered. Added to this is a growing sense of guilt with the feeling that his children are failing and not moving forward in life. While one is unemployed, uses cannabis and spends his time playing video games, the other has no education and has fallen into delinquency. He is desperate about their life trajectory, and experiences the failure of his attempts to help them as a personal failure and proof of his incompetence.

He says he always needs someone behind him to support him and tell him that he is on the right track. Although he feels that his situation is objectively difficult and generates suffering, he questions the fact that his psychological state can contribute to make him increase the difficulties of the situation at the expense of possible improvements.

After this first interview, a set of additional information gathering devices are proposed to Bob in preparation for a second interview. In order to refine the diagnostic elements, it is suggested that he answers a self-administered online questionnaire adapted from Structured Clinical Interview for DSM (SCID), the Screening Assessment for Guiding Evaluation-Self-Report-SAGE-SR, [54], which makes it possible to obtain elements on the connection of the symptoms described to the various syndromic categories of the DSM. The multidisciplinary team has a platform giving them access to different databases of connected objects, software databases, health parameters from which they can select what is relevant to Bob's situation. This is how he is offered a dedicated application integrating the monitoring of his sleep, his daily activities, his heart rate, his eating habits and his emotional state in the different situations of daily life [55]. An activity sensor and a GPS in the form of a connected watch not identifiable as a medical device, and usable in everyday life makes it possible to collect objective physiological and behavioral data. A home automation sensor located in the home rooms collects environmental data (temperature, noise, light, air quality, presence in the room). An analysis of the use of its means of communication (telephone calls, sms, e-mails, social networks) is also implemented [48,56,57].

Three weeks later, the multidisciplinary team thus has a set of psychopathological data based on a validated self-evaluation instrument and objective behavioral and environmental data to supplement the first clinical and paraclinical examinations. These data can be used during a second more in-depth teleconsultation (Figure 1, step 2) designed to co-construct with Bob an idiographic model of the main somatic, psychological, relational and situational factors involved in the genesis and maintenance of his problems [26] and to propose adapted interventions. The data collected confirms the existence in Bob of several elements relating to anxious; depressive and adaptation disorders constellations; inefficient coping modalities in different professional situations and in interactions with his children; indicators of difficulties in managing different stressors (intense heart rate acceleration, strong emotional variations) in particular public speaking and interaction with customers; sleep difficulties and late night insomnia associated with intense snoring episodes and breathing interruptions suggestive of sleep apnea; as well as an excessively low level of physical activity and sedentary lifestyle associated with a limited scope of life, with no outside activity [58]. Diet monitoring shows a very unbalanced diet, with an excessive load of high glycaemic index foods, too little protein and excessive alcohol consumption, especially when returning home and in the late evening. As for the analysis of communications, it shows that Bob has cut himself off from his social and friendly network, and that his communications are limited to exchanges with his wife, most often associated with episodes of anxiety during his travels and professional activities. Automated analysis of his text messages shows significant use of a lexical field associated with anxiety, exhaustion and panic [59], while automatic analysis of his voice during communications also shows anxious elements in communications with his wife, and elements of depression and emotional restriction in his exchanges with clients and colleagues [60].

During this second consultation, the multidisciplinary team and Bob can study together the different intervention modalities available in the system bases in order to choose those that seem best adapted to Bob's needs and current functioning. This is why a tele-psychotherapy is proposed to him in order to address what he has identified as one of the important causes in his current difficulties, namely the badly managed mourning of his mother's death. Added to this are various software modules designed to enable him to better manage his insomnia problems (e.g., Sleepcare [61]), to gradually overcome his difficulties in making projects, his aboulia and his anhedonia thanks to e-CBT protocols (e.g., MoodHacker [62] and Get Happy [63]), improve psychological flexibility, regulate emotions and guilt more effectively, and identify what is really important to him in life in order to act in this direction using elements of Acceptance-Based Therapy (e.g., ACTsmart [64] and ibobbly [65]), to facilitate his organization and improve his attentional control, especially at work in order to help him better cope with psychological pressure problems (e.g., LivingSMART [66]), to support his sense of self-esteem, acceptance of his personal identity, in order to improve his self-image, especially in relationships with others, particularly with his children, customers and superiors (e.g., SuperBetter [67]). Bob's sleep difficulties are the subject of several synergistic interventions. The connection of the e-mental health device to home automation devices allows a dynamic management of the environment adapted to the sleep profile. The system thus acts on the inside home temperature, progressively reduces the intensity of light and the quantity of blue light diffused by the screens and the connected bulbs, proposes adapted relaxation and sleep preparation interventions at appropriate times, and suggests an appropriate distribution of food categories for dinner. Sleeping or sleep promoting products having been prescribed, the device proposes in a way adapted to the profile of the past day and that of the day to come the use of one type of molecule or another (phytotherapeutic complexes, melatonin, doxylamine, zolpidem, zopiclone . . .) and manages the duration of treatment and the doses used, taking into account the risks of adverse effects and the development of an addiction. The severity of Bob's depressive state (30 on Hamilton scale) also justifies the use of an adapted pharmacological treatment [6] in the form of an Selective serotonin reuptake inhibitors (SSRI). A specific follow-up of the health parameters theoretically affected by this treatment is set up in the system allowing an objectification of both its potential therapeutic and undesirable effects.

The device also includes the data collection modules used from the first session, thus allowing Bob a longitudinal follow-up and a better knowledge of his own functioning. This "technologically" increased reflexivity is likely by itself to have effects on the improvement of functioning and symptoms [68]. There are also psycho-educational elements on mental health management (e.g., MyCompass [69]), which can be communicated in direct relation to the context and events, thus improving their relevance and effective integration in practice. In order to avoid the frequently observed decrease in the rate of device use and compliance [70,71], the proposed EMA/EMI applications are designed with particular attention to user commitment and motivation, in particular through the use of gamification [72]. The ergonomics and graphics of the interface allowing interaction with the system are designed to promote a pleasant navigation experience for both patients and clinicians. Overall, the device technologically supports new possibilities of empowerment, self-management and empowerment for Bob, in particular through the understanding of the fine mechanisms that generate and maintain problematic functioning and the operational possibility given to the patient to act on them [48].

The individual data collected by the device are also used to create a data warehouse that serves as a basis for collective analyses of multiple patients. Each element of this database is a localized observation or intervention data collected at time t in a patient p. Tuple analysis {d, p, t} allows both longitudinal intra-individual analyses and inter-individual comparisons or aggregated group analyses, and allows the progressive constitution of a nomothetic and idiographic knowledge base accessible to clinicians and patients to evaluate the effects in real situations of the therapeutic actions implemented.

10. Conclusions

The use of connected mobile technologies in the field of mental health has developed particularly rapidly in recent years. Although many applications and connected objects continue to be offered directly by companies, or even individual developers, without their reliability, effectiveness or undesirable effects being scientifically evaluated, it is nevertheless clear that the "technological gadget" stage has now largely passed. Numerous observational and experimental studies of m-health devices have indeed already been conducted, and show statistically and clinically significant effect sizes on many mental disorders and mental health problems.

The central argument of this article is that in order to make the most of these technological innovations, epistemological innovations in the design of mental disorders and therapies are indispensable. The approaches to disorders that we currently use, based on categorical definitions that hypothesize latent causes, were designed with the observational and interventional tools that were available when these conceptualizations emerged. If only for strictly logistical reasons (impossibility of monitoring, observing and intervening on patients every second of their lives), these approaches were based on macroscopic conceptualizations of the disorders associated with specific interventions.

New connected mobile technologies offer the possibility of ubiquitous micro-observations and micro-interventions. This technological finesse of observation and intervention must be matched by an equivalent finesse in models of understanding of disorders and treatments. Our hypothesis is that "symptom network" approaches, such as the one proposed by Borsboom [3] and many other researchers following him open up a particularly interesting avenue for this.

Research and developments on the synergies between new technological approaches to m-health and new epistemological approaches to mental disorders, while promising, remain at an embryonic stage. To enable them to develop effectively, it is also essential to add methodological innovations to them, particularly with regard to effectiveness evaluation mechanisms [73]. The current gold standard of the randomized controlled trial [74] forces experimental studies to limit themselves to evaluating the effect of a single intervention on a single disease entity when, as we have shown, the future lies much more in the possibility of coordinating multiple fine interventions on multiple fine mechanisms.

It is therefore necessary to coordinate efforts now in technological, epistemological and methodological innovations to enable implementation of real-time assessments that have the potential to guide clinical decision toward more appropriate and targeted therapeutic interventions tailored to each individual case, as well as ensuring the rigorous monitoring of standard treatment strategies classically proposed in routine clinical practice thereby improving overall effectiveness and adherence, and thus changing relapsing course and poor prognosis of mental disorders.

Author Contributions: X.B., M.M. and P.C. had contributed to design and to write the paper.

Funding: This research received no external funding.

Acknowledgments: We are grateful for the many exchanges with Karim N'Diaye and Luc Mallet that have greatly contributed to the development of the ideas presented in this text.

Conflicts of Interest: The authors declare no conflict of interest.

References

1. Brun, C.; Demazeux, S.; Vittorio, P.D.; Gonon, F.; Gorry, P.; Konsman, J.P.; Lung, F.; Lung, Y.; Minard, M.; Montalban, M.; et al. La construction des catégories diagnostiques de maladie mentale. *Rev La Régulation* **2015**, *17*. [CrossRef]

2. Demazeux, S. *Qu'est-ce Que Le DSM? Genèse et Transformations de la Bible Américaine de La Psychiatrie*; Ithaque: Paris, France, 2013.

3. Borsboom, D. A network theory of mental disorders. *World Psychiatry* **2017**, *16*, 5–13. [CrossRef] [PubMed]

4. Wampold, B.E. *The Great Psychotherapy Debate: Models, Methods, and Findings*; Routledge: Abingdon, UK, 2013; 280p.

5. Gaynes, B.N.; Warden, D.; Trivedi, M.H.; Wisniewski, S.R.; Fava, M.; Rush, A.J. What did STAR* D teach us? Results from a large-scale, practical, clinical trial for patients with depression. *Psychiatr. Serv.* **2009**, *60*, 1439–1445. [CrossRef] [PubMed]

6. Kirsch, I.; Deacon, B.J.; Huedo-Medina, T.B.; Scoboria, A.; Moore, T.J.; Johnson, B.T. Initial severity and antidepressant benefits: A meta-analysis of data submitted to the Food and Drug Administration. *PLoS Med. Public Libr. Sci.* **2008**, *5*, e45. [CrossRef] [PubMed]

7. Borsboom, D. Psychometric Perspectives on Diagnostic Systems. *J. Clin. Psychol.* **2008**, *64*, 1089–1108. [CrossRef] [PubMed]

8. Cramer, A.O.J.; Waldorp, L.J.; van der Maas, H.L.J.; Borsboom, D. Comorbidity: A network perspective. *Behav. Brain Sci.* **2010**, *33*, 137–193. [CrossRef] [PubMed]
9. Fried, E.I.; van Borkulo, C.D.; Cramer, A.O.J.; Boschloo, L.; Schoevers, R.A.; Borsboom, D. Mental disorders as networks of problems: A review of recent insights. *Soc. Psychiatry Psychiatr. Epidemiol.* **2016**, *58*, 7250–7257. [CrossRef] [PubMed]
10. Borsboom, D.; Cramer, A.O.J.J. Network Analysis: An Integrative Approach to the Structure of Psychopathology. *Annu. Rev. Clin. Psychol.* **2013**, *9*, 91–121. [CrossRef] [PubMed]
11. Béliard, A.; Eideliman, J.-S. Aux frontières du handicap psychique: Genèse et usages des catégories médico-administratives. *Revue Française Des Affaires Sociales* **2010**, *1*, 99–117.
12. Strand, M.; Gammon, D.; Ruland, C.M. Transitions from biomedical to recovery-oriented practices in mental health: A scoping review to explore the role of Internet-based interventions. *BMC Health Serv. Res.* **2017**, *17*, 1–14. [CrossRef] [PubMed]
13. Morgiève, M.; Ung, Y.; Céline Gehamy, X.B. Diminuer l' impact des troubles obsessionnels compulsifs par des modifications de l' environnement physique Une étude de preuve de concept. *Psychiatr. Sci. Hum. Neurosci.* **2016**, *14*, 43–63.
14. Gehamy, C.; Morgiève, M.; Briffault, X. Design participatif en santé mentale: Le cas des troubles obsessionnels compulsifs. *Sci. Du Des.* **2017**, *2*, 80–91.
15. Nelson, B.; McGorry, P.D.; Wichers, M.; Wigman, J.T.W.; Hartmann, J.A. Moving from static to dynamic models of the onset of mental disorder: A review. *JAMA Psychiatry* **2017**, *74*, 528–534. [CrossRef] [PubMed]
16. Wichers, M. The dynamic nature of depression: A new micro-level perspective of mental disorder that meets current challenges. *Psychol. Med.* **2014**, *44*, 1349–1360. [CrossRef] [PubMed]
17. Cramer, A.O.J.; van Borkulo, C.D.; Giltay, E.J.; van der Maas, H.L.J.; Kendler, K.S.; Scheffer, M.; Borsboom, D. Major depression as a complex dynamic system. *PLoS ONE* **2016**, *11*, e0167490. [CrossRef] [PubMed]
18. Plagnol, A.; Pachoud, B. Capacité fonctionnelle et fonctionnement en situation réelle. In *Handicap Psychique: Questions Vives*; Éditions Érès: Toulouse, France, 2016; pp. 193–213.
19. Harari, G.M.; Müller, S.R.; Aung, M.S.; Rentfrow, P.J. Smartphone sensing methods for studying behavior in everyday life. *Curr. Opin. Behav. Sci.* **2017**, *18*, 83–90. [CrossRef]
20. Harari, G.M.; Lane, N.D.; Wang, R.; Crosier, B.S.; Campbell, A.T.; Gosling, S.D. Using Smartphones to Collect Behavioral Data in Psychological Science. *Perspect. Psychol. Sci.* **2016**, *11*, 838–854. [CrossRef] [PubMed]
21. Adams, Z.; McClure, E.A.; Gray, K.M.; Danielson, C.K.; Treiber, F.A.; Ruggiero, K.J. Mobile devices for the remote acquisition of physiological and behavioral biomarkers in psychiatric clinical research. *J. Psychiatr. Res.* **2017**, *85*, 1–14. [CrossRef] [PubMed]
22. Firth, J.; Torous, J.; Yung, A.R. Ecological momentary assessment and beyond: The rising interest in e-mental health research. *J. Psychiatr. Res.* **2016**, *80*, 3–4. [CrossRef] [PubMed]
23. García, C.G.; Meana-Llorián, D.; G-Bustelo, B.C.P.; Lovelle, J.M.C. A review about Smart Objects, Sensors, and Actuators. *Int. J. Interact. Multimedia Artif. Intell.* **2017**, *4*, 7–10. [CrossRef]
24. Briffault, X.; Morgiève, M. Anticiper les usages et les conséquences des technologies connectées en santé mentale. Une étude de « cas fictif » *J. Médecin Légale* **2018**, in press.
25. Van Borkulo, C.; Boschloo, L.; Borsboom, D.; Penninx, B.W.J.H.; Waldorp, L.J.; Schoevers, R.A. Association of Symptom Network Structure With the Course of Depression. *JAMA Psychiatry* **2015**, *72*, 1219. [CrossRef] [PubMed]
26. Schiepek, G.K.; Stoger-Schmidinger, B.; Aichhorn, W.; Scholler, H.; Aas, B. Systemic case formulation, individualized process monitoring, and state dynamics in a case of dissociative identity disorder. *Front. Psychol.* **2016**, *7*, 1–11. [CrossRef] [PubMed]
27. Kreuze, E.; Jenkins, C.; Gregoski, M.; York, J.; Mueller, M.; Lamis, D.A.; Ruggiero, K.J. Technology-enhanced suicide prevention interventions: A systematic review of the current state of the science. *J. Telemed. Telecare* **2016**, *23*, 1–13.
28. Li, X.; Dunn, J.; Salins, D.; Zhou, G.; Zhou, W.; Schüssler-Fiorenza Rose, S.M.; Perelman, D.; Colbert, E.; Runge, R.; Rego, S.; et al. Digital Health: Tracking Physiomes and Activity Using Wearable Biosensors Reveals Useful Health-Related Information. *PLoS Biol.* **2017**, *15*, e2001402. [CrossRef] [PubMed]
29. Callan, J.A.; Wright, J.; Siegle, G.J.; Howland, R.H.; Kepler, B.B. Use of Computer and Mobile Technologies in the Treatment of Depression. *Arch. Psychiatr. Nurs.* **2016**, *31*, 311–318. [CrossRef] [PubMed]

30. Meredith, S.E.; Alessi, S.M.; Petry, N.M. Smartphone applications to reduce alcohol consumption and help patients with alcohol use disorder: A state-of-the-art review. *Adv. Health Care Technol.* **2016**, *1*, 47–54.

31. Bakker, D.; Kazantzis, N.; Rickwood, D.; Rickard, N. Mental Health Smartphone Apps: Review and Evidence-Based Recommendations for Future Developments. *JMIR Ment. Health* **2016**, *3*, e7. [CrossRef] [PubMed]

32. Coulon, S.M.; Monroe, C.M.; West, D.S. A Systematic, Multi-domain Review of Mobile Smartphone Apps for Evidence-Based Stress Management. *Am. J. Prev. Med.* **2016**, *51*, 95–105. [CrossRef] [PubMed]

33. Van Ameringen, M.; Turna, J.; Khalesi, Z.; Pullia, K.; Patterson, B. There is an app for that! The current state of mobile applications (apps) for DSM-5 obsessive-compulsive disorder, posttraumatic stress disorder, anxiety and mood disorders. *Depress Anxiety* **2017**, 1–14. [CrossRef] [PubMed]

34. Huguet, A.; Rao, S.; McGrath, P.J.; Wozney, L.; Wheaton, M.; Conrod, J.; Rozario, S. A systematic review of cognitive behavioral therapy and behavioral activation apps for depression. *PLoS ONE* **2016**, *11*, 1–19. [CrossRef] [PubMed]

35. Donker, T.; Blankers, M.; Hedman, E.; Ljotsson, B.; Petrie, K.; Christensen, H. Economic evaluations of Internet interventions for mental health: A systematic review. *Psychol. Med.* **2015**, *45*, 3357–3376. [CrossRef] [PubMed]

36. Paganini, S.; Teigelkötter, W.; Buntrock, C.; Baumeister, H. Economic evaluations of internet- and mobile-based interventions for the treatment and prevention of depression: A systematic review. *J. Affect. Disord.* **2018**, *225*, 733–755. [CrossRef] [PubMed]

37. Iribarren, S.J.; Cato, K.; Falzon, L.; Stone, P.W. What is the economic evidence for mHealth? A systematic review of economic evaluations of mHealth solutions. *PLoS ONE* **2017**, *12*, 1–20. [CrossRef] [PubMed]

38. Berrouiguet, S.; Perez-Rodriguez, M.M.; Larsen, M.; Baca-García, E.; Courtet, P.; Oquendo, M. From eHealth to iHealth: Transition to Participatory and Personalized Medicine in Mental Health. *J. Med. Internet Res.* **2018**, *20*, e2. [CrossRef] [PubMed]

39. Briffault, X. Singularisations, contextualisations, interconnexions. *Perspect. Psy* **2017**, *56*, 133–141. [CrossRef]

40. Briffault, X.; Morgiève, M. François Vatel se serait-il suicidé s'il avait eu un smartphone? Potentiels de soin et conséquences épistémologiques des technologies mobiles en santé mentale. Mental healthcare potentials and epistemological consequences of mobile technologies. *PSN* **2017**, *15*, 47–70.

41. Shen, N.; Levitan, M.-J.; Johnson, A.; Bender, J.L.; Hamilton-Page, M.; Jadad, A.A.R.; Wiljer, D. Finding a depression app: A review and content analysis of the depression app marketplace. *JMIR mHealth uHealth* **2015**, *3*, e16. [CrossRef] [PubMed]

42. Trajković, G.; Starčević, V.; Latas, M.; Leštarević, M.; Ille, T.; Bukumirić, Z.; Marinković, J. Reliability of the Hamilton Rating Scale for Depression: A meta-analysis over a period of 49 years. *Neuropsychopharmacology* **2011**, *189*, 1–9. [CrossRef] [PubMed]

43. Beck, A.T.; Steer, R.A.; Carbin, M.G. Psychometric properties of the Beck Depression Inventory: Twenty-five years of evaluation. *Clin. Psychol. Rev.* **1988**, *8*, 77–100. [CrossRef]

44. Kroenke, K.; Spitzer, R.L. The PHQ-9: A New Depression Diagnostic and Severity Measure. *Psychiatr. Ann.* **2002**, *32*, 509–515. [CrossRef]

45. Ahearn, E.P. The use of visual analog scales in mood disorders: A critical review. *J. Psychiatr. Res.* **1997**, *31*, 569–579. [CrossRef]

46. Saeb, S.; Zhang, M.; Karr, C.J.; Schueller, S.M.; Corden, M.E.; Kording, K.P.; Mohr, D.C. Mobile phone sensor correlates of depressive symptom severity in daily-life behavior: An exploratory study. *J. Med. Internet Res.* **2015**, *17*, 1–11. [CrossRef] [PubMed]

47. Valstar, M.; Schuller, B.; Smith, K.; Eyben, F.; Jiang, B.; Bilakhia, S.; Schnieder, S.; Cowie, R.; Pantic, M. AVEC 2013: The continuous audio/visual emotion and depression recognition challenge. In Proceedings of the 3rd ACM International Workshop on Audio/Visual Emotion Challenge, Barcelona, Spain, 21–25 October 2013; pp. 3–10.

48. Dogan, E.; Sander, C.; Wagner, X.; Hegerl, U.; Kohls, E. Smartphone-Based monitoring of objective and subjective data in affective disorders: Where are we and where are we going? Systematic review. *J. Med. Internet Res.* **2017**, *19*, e262. [CrossRef] [PubMed]

49. Wang, W.; Li, Y.; Huang, Y.; Liu, H.; Zhang, T. A Method for Identifying the Mood States of Social Network Users Based on Cyber Psychometrics. *Future Internet* **2017**, *9*, 22. [CrossRef]

50. Mohr, D.C.; Lyon, A.R.; Lattie, E.G.; Reddy, M.; Schueller, S.M. Accelerating digital mental health research from early design and creation to successful implementation and sustainment. *J. Med. Internet Res.* **2017**, *19*, 1–14. [CrossRef] [PubMed]

51. Jimenez, P.; Bregenzer, A. Integration of eHealth Tools in the Process of Workplace Health Promotion: Proposal for Design and Implementation. *J. Med. Internet Res.* **2018**, *20*, e65. [CrossRef] [PubMed]

52. Derks, Y.P.M.J.; De Visser, T.; Bohlmeijer, E.T.; Noordzij, M.L. MHealth in Mental Health: How to efficiently and scientifically create an ambulatory biofeedback e-coaching app for patients with borderline personality disorder. *Int. J. Hum. Fact. Ergon.* **2017**, *5*, 61–92. [CrossRef]

53. Vallury, K.D.; Jones, M.; Oosterbroek, C. Computerized cognitive behavior therapy for anxiety and depression in rural areas: A systematic review. *J. Med. Internet Res.* **2015**, *17*, e139. [CrossRef] [PubMed]

54. Brodey, B.; Purcell, S.E.; Rhea, K.; Maier, P.; First, M.; Zweede, L.; Sinisterra, M.; Nunn, M.B.; Austin, M.P.; Brodey, I.S. Rapid and Accurate Behavioral Health Diagnostic Screening: Initial Validation Study of a Web-Based, Self-Report Tool (the SAGE-SR). *J. Med. Internet Res.* **2018**, *20*, e108. [CrossRef] [PubMed]

55. Faurholt-Jepsen, M.; Vinberg, M.; Frost, M.; Christensen, E.M.; Bardram, J.; Kessing, L.V. Daily electronic monitoring of subjective and objective measures of illness activity in bipolar disorder using smartphones–the MONARCA II trial protocol: A randomized controlled single-blind parallel-group trial. *BMC Psychiatry* **2014**, *14*, 309. [CrossRef] [PubMed]

56. Grünerbl, A.; Muaremi, A.; Osmani, V.; Bahle, G.; Oehler, S.; Tröster, G.; Mayora, O.; Haring, C.; Lukowicz, P. Smartphone-based recognition of states and state changes in bipolar disorder patients. *IEEE J. Biomed. Health Inform.* **2015**, *19*, 140–148. [CrossRef] [PubMed]

57. Alvarez-Lozano, J.; Osmani, V.; Mayora, O.; Frost, M.; Bardram, J.; Faurholt-Jepsen, M.; Kessing, L.V. Tell me your apps and I will tell you your mood: Correlation of apps usage with bipolar disorder state. In Proceedings of the 7th International Conference on PErvasive Technologies Related to Assistive Environments, Rhodes, Greece, 27–30 May 2014; p. 19.

58. Grüenerbl, A.; Osmani, V.; Bahle, G.; Carrasco, J.C.; Oehler, S.; Mayora, O.; Haring, C.; Lukowicz, P. Using smart phone mobility traces for the diagnosis of depressive and manic episodes in bipolar patients. In Proceedings of the 5th Augmented Human International Conference, Kobe, Japan, 2014; pp. 1–8.

59. Al-Mosaiwi, M.; Johnstone, T. In an Absolute State: Elevated Use of Absolutist Words Is a Marker Specific to Anxiety, Depression, and Suicidal Ideation. *Clin. Psychol. Sci.* **2018**. [CrossRef]

60. Hashim, N.W.; Wilkes, M.; Salomon, R.; Meggs, J.; France, D.J. Evaluation of Voice Acoustics as Predictors of Clinical Depression Scores. *J. Voice* **2017**, *31*, 256.e1–256.e6. [CrossRef] [PubMed]

61. Horsch, C.H.G.; Lancee, J.; Griffioen-Both, F.; Spruit, S.; Fitrianie, S.; Neerincx, M.A.; Beun, R.J.; Brinkman, W.P. Mobile phone-delivered cognitive behavioral therapy for insomnia: A randomized waitlist controlled trial. *J. Med. Internet Res.* **2017**, *19*, e70. [CrossRef] [PubMed]

62. Birney, A.J.; Gunn, R.; Russell, J.K.; Ary, D.V. MoodHacker mobile web app with email for adults to self-manage mild-to-moderate depression: Randomized controlled trial. *JMIR mHealth uHealth* **2016**, *4*, e8. [CrossRef] [PubMed]

63. Watts, S.; Mackenzie, A.; Thomas, C.; Griskaitis, A.; Mewton, L.; Williams, A.; Andrews, G. CBT for depression: A pilot RCT comparing mobile phone vs. computer. *BMC Psychiatry* **2013**, *13*, 49. [CrossRef] [PubMed]

64. Ivanova, E.; Lindner, P.; Ly, K.H.; Dahlin, M.; Vernmark, K.; Andersson, G.; Carlbring, P. Guided and unguided Acceptance and Commitment Therapy for social anxiety disorder and/or panic disorder provided via the Internet and a smartphone application: A randomized controlled trial. *J. Anxiety Disord.* **2016**, *44*, 27–35. [CrossRef] [PubMed]

65. Tighe, J.; Shand, F.; Ridani, R.; Mackinnon, A.; De La Mata, N.; Christensen, H. Ibobbly mobile health intervention for suicide prevention in Australian Indigenous youth: A pilot randomised controlled trial. *BMJ Open* **2017**, *7*, e013518. [CrossRef] [PubMed]

66. Moëll, B.; Kollberg, L.; Nasri, B.; Lindefors, N.; Kaldo, V. Living SMART—A randomized controlled trial of a guided online course teaching adults with ADHD or sub-clinical ADHD to use smartphones to structure their everyday life. *Internet Interv.* **2015**, *2*, 24–31. [CrossRef]

67. Roepke, A.M.; Jaffee, S.R.; Riffle, O.M.; McGonigal, J.; Broome, R.; Maxwell, B. Randomized controlled trial of SuperBetter, a smartphone-based/Internet-based self-help tool to reduce depressive symptoms. *Games Health J.* **2015**, *4*, 235–246. [CrossRef] [PubMed]

68. Firth, J.; Torous, J.; Nicholas, J.; Carney, R.; Pratap, A.; Rosenbaum, S.; Sarris, J. The efficacy of smartphone-based mental health interventions for depressive symptoms: A meta-analysis of randomized controlled trials. *World Psychiatry* **2017**, *16*, 287–298. [CrossRef] [PubMed]

69. Proudfoot, J.; Clarke, J.; Birch, M.-R.; Whitton, A.E.; Parker, G.; Manicavasagar, V.; Harrison, V.; Christensen, H.; Hadzi-Pavlovic, D. Impact of a mobile phone and web program on symptom and functional outcomes for people with mild-to-moderate depression, anxiety and stress: A randomised controlled trial. *BMC Psychiatry* **2013**, *13*, 312. [CrossRef] [PubMed]

70. Burns, M.N.; Begale, M.; Duffecy, J.; Gergle, D.; Karr, C.J.; Giangrande, E.; Mohr, D.C. Harnessing context sensing to develop a mobile intervention for depression. *J. Med. Internet Res.* **2011**, *13*, e55. [CrossRef] [PubMed]

71. Anguera, J.A.; Jordan, J.T.; Castaneda, D.; Gazzaley, A.; Areán, P.A. Conducting a fully mobile and randomised clinical trial for depression: Access, engagement and expense. *BMJ Innov. BMJ Spec. J.* **2016**, *2*, 14–21. [CrossRef] [PubMed]

72. Comello, M.L.G.; Qian, X.; Deal, A.M.; Ribisl, K.M.; Linnan, L.A.; Tate, D.F. Impact of game-inspired infographics on user engagement and information processing in an eHealth program. *J. Med. Internet Res.* **2016**, *18*, e237. [CrossRef] [PubMed]

73. Teng, P.; Bateman, N.W.; Darcy, K.M.; Hamilton, C.A.; Maxwell, G.L.; Bakkenist, C.J. Realizing the Potential of Mobile Mental Health: New Methods for New Data in Psychiatry. *Curr. Psychiatry Rep.* **2015**, *17*, 13.

74. Porter, R.; Frampton, C.; Joyce, P.R.; Mulder, R.T. Randomized controlled trials in psychiatry. Part 1: Methodology and critical evaluation. *Aust. N. Z. J. Psychiatry* **2003**, *37*, 257–264. [CrossRef] [PubMed]

brain sciences

MDPI

Review

Individualized Immunological Data for Precise Classification of OCD Patients

Hugues Lamothe [1,2,3], **Jean-Marc Baleyte** [1,3], **Pauline Smith** [2], **Antoine Pelissolo** [3,4,5] **and Luc Mallet** [2,3,4,6,*]

1   Centre Hospitalier Intercommunal de Créteil, 94000 Créteil, France; lamothehugues@gmail.com (H.L.); jean-marc.baleyte@chicreteil.fr (J.-M.B.)
2   Institut du Cerveau et de la Moelle Epinière, Sorbonne Universités, UPMC Univ Paris 06, CNRS, INSERM, 75013 Paris, France; pauline.hh.smith@gmail.com
3   Fondation FondaMental, 94000 Créteil, France; a.pelissolo@gmail.com
4   Assistance Publique-Hôpitaux de Paris, Pôle de Psychiatrie, Hôpitaux Universitaires Henri Mondor—Albert Chenevier, Université Paris-Est Créteil, 94000 Créteil, France
5   INSERM, U955, Team 15, 94000 Créteil, France
6   Department of Mental Health and Psychiatry, Global Health Institute, University of Geneva, 1202 Geneva, Switzerland
*   Correspondence: luc.mallet@inserm.fr; Tel.: +33-157-274-393

Received: 16 May 2018; Accepted: 3 August 2018; Published: 9 August 2018

Abstract: Obsessive–compulsive disorder (OCD) affects about 2% of the general population, for which several etiological factors were identified. Important among these is immunological dysfunction. This review aims to show how immunology can inform specific etiological factors, and how distinguishing between these etiologies is important from a personalized treatment perspective. We found discrepancies concerning cytokines, raising the hypothesis of specific immunological etiological factors. Antibody studies support the existence of a potential autoimmune etiological factor. Infections may also provoke OCD symptoms, and therefore, could be considered as specific etiological factors with specific immunological impairments. Finally, we underline the importance of distinguishing between different etiological factors since some specific treatments already exist in the context of immunological factors for the improvement of classic treatments.

Keywords: psychiatry; OCD; obsessive–compulsive disorder; Tourette syndrome; immunology; cytokines; pediatric autoimmune neuropsychological disorders associated with streptococcal infection (PANDAS); pediatric acute-onset neuropsychiatric syndrome (PANS); *Toxoplasma gondii*; *Streptococcus pyogenes*

1. Introduction

Obsessive–compulsive disorder (OCD) is a major disabling disorder affecting about 2% of the population, and it incurs significant mental health costs [1]. The Diagnostic and Statistical Manual of Mental Disorders, Fifth Edition (DSM-5) defines OCD as comprising two major symptoms: obsessions (i.e., intrusive thoughts or mental images) and compulsions (i.e., repetitive movements or mental acts produced by the patient in response to obsessional thoughts, in order to decrease anxiety) [2].

Several hypotheses exist regarding the physiological basis of OCD with dysfunction of brain circuits involving the limbic cortex and basal ganglia being at the core of the disorder [1,3]. Indeed, several imaging studies found hyperactivity of the orbito-frontal cortex and anterior cingulate cortex [4], and effective treatments for severe forms of OCD act directly on these circuits [5,6]. Some authors proposed hypotheses involving dysfunction of microcircuits within these limbic loops [7]. However, hypotheses constructed to explain the underlying pathology of the disorders make no reference as to the origin of the dysfunctions.

An underlying genetic process could play an important role in the etiology of dysfunctional circuitry. Several genes were found through genomic association studies [1,8]. Among the genes implicated in OCD, dopamine-, glutamate-, or serotonin-related genes are the most studied [1,3], although they are not the only ones to be involved in OCD. A recent study aimed to identify rare de novo mutations based on exomes from 20 OCD parent–child trios patients [9]. Protein mutations were found in developmental and immunological pathways, such as transforming growth factor beta (TGFβ) or complements. These results differ from the usual neurotransmitter gene mutations [9], and they provide arguments for immunological factors in OCD etiology. Furthermore, another study found a significant enrichment of the human leukocyte antigen/antigen D-related 4 (HLA-DR4) serotype allele in OCD patients [10]. According to the results of these genetic studies, abnormalities in immunological mechanisms could lead to OCD, and some specific mechanisms such as microglial dysfunction [11] or autoimmune processes [12] were hypothesized. Furthermore, not only can genes disrupt the immune system through mutations, but the environment can also influence it, through infections for example, subsequently leading to OCD, even with no genetic predisposition [13]. This question of a possible infectious etiology was also suspected in other psychiatric disorders [14].

As many as 25–40% of OCD patients are resistant to classical therapies, such as serotonin recapture inhibitors and cognitive behavioral therapy, and remain so despite advances in OCD treatment such as deep brain stimulation [15–17]. Any progress toward a better understanding of the biological basis could provide some solutions for resistant OCD patients. Hence, if a specific cluster of OCD patients with immunological dysfunctions could be determined, some specific treatment could emerge for them and resolve the enduring resistance problem for a minority of patients [18]. The aim of this review was, thus, to summarize the existing immunological data in OCD to show a possible immunological etiological factor in OCD that could be distinct from other factors (e.g., purely genetic OCD), and then, to raise the possibility of a more personalized and effective treatment.

2. Method

Our review used the PubMed database. We selected clinical human papers (English language) relevant to the human immunological field concerning OCD. The relevance of an article was based on the abstract published in PubMed. We did not restrain the period of search and reviewed all PubMed results.

Our exclusion criteria were as follows: case report format, small descriptive case series format, commentary format, review format, animal experiments, neurocognitive studies, control group with other psychiatric conditions, absence of OCD data (for example, studies looking for pediatric autoimmune neuropsychological disorders associated with streptococcal infection (PANDAS) etiology only in Tourette's patients were excluded). Furthermore, therapeutic trials were not selected when these were not targeted by the search terms. For example, when looking for cytokine impairment in OCD through "cytokines AND (OCD OR "obsessive compulsive disorder")" search terms, some therapeutic trials found were not selected.

When the same paper was found with different search terms, we specify this fact in the tables below. For example, when an article was found with both "cytokines AND (OCD OR "obsessive compulsive disorder")" and "antibody AND (OCD OR "obsessive compulsive disorder")" search terms, we detailed the article only in the first table section (here, the "cytokines AND (OCD OR "obsessive compulsive disorder")" table section); in the second table section (here, the "antibody AND (OCD OR "obsessive compulsive disorder")" table section), we only mentioned the article and directed the reader to the first table section for details.

The following terms were reviewed:

cytokines AND (OCD OR "obsessive-compulsive disorder") (67 papers, 22 included); antibody AND (OCD OR "obsessive-compulsive disorder") (163 papers, 33 included); anti-brain antibody and (OCD OR "obsessive-compulsive disorder") (6 papers, 3 included); ABGA AND (OCD OR "obsessive-compulsive disorder") (7 papers, 2 included); "white blood cells" AND (OCD OR "obsessive-compulsive disorder") (1 paper, 1 included); lymphocyte AND ("obsessive-compulsive disorder" OR OCD) (40 papers, 15 included); monocytes AND ("obsessive-compulsive disorder" OR OCD) (6 papers, 4 included); "NK cells" AND ("obsessive-compulsive disorder" OR OCD) (3 papers, 2 included); infection AND (OCD OR "obsessive-compulsive disorder") (243 papers, 9 included); Lyme AND (OCD OR "obsessive-compulsive disorder") (5 papers, 1 included); streptococcus AND (OCD OR "obsessive-compulsive disorder") (137 papers, 22 included); toxoplasma (OCD OR "obsessive-compulsive disorder") (8 papers, 4 included); (PANDAS OR PANS) AND treatment AND (OCD OR "obsessive-compulsive disorder" OR tic OR Tourette) (111 papers, 21 included); NSAID and (OCD OR "obsessive–compulsive disorder") (14 articles, 4 included); "anti-inflammatory" and (OCD OR "obsessive–compulsive disorder") (21 papers, 5 included); minocycline and (OCD OR "obsessive–compulsive disorder") (6 papers, 2 included); N-acetylcysteine and (OCD OR "obsessive–compulsive disorder") (27 papers, 5 included).

The aim of the article was to describe and discuss the potential role of immunological factors in OCD etiology. Hence, even if the method was a systematic one, we wrote our article as a qualitative review to make it easier to read and understand. However, the articles cited in the text are referenced in the tables below, and, when contradictions occurred between articles, this is mentioned and reviewed qualitatively in the text.

3. Immunological Changes in OCD

3.1. Cytokines

Cytokines are molecules that allow communication between immune cells, or between immune cells and non-immune cells [19]. Studying cytokines can help us understand the mechanism and pathways of potential immunological disruption in OCD. The first studies on cytokine variations in OCD patients included very few patients and were negative except for a positive and significant correlation between IL-6 (interleukin-6) or soluble IL-6 receptor plasma levels and severity of compulsive behaviors [20,21] (Table 1). Further studies [22–30] were carried out enabling a meta-analysis [31] (Table 1), which found decreased IL-1β levels and decreased TNFα (tumor necrosis factor α) levels in non-depressed OCD patients (but not in OCD patients with possible comorbid depression), and increased IL-6 levels in adult medication-free OCD patients (but not in OCD children with possible medication use) compared to controls. More recently, discrepancies were found with this previous meta-analysis concerning TNF-α with increased levels in OCD patients [32–34] (Table 1). Despite these discrepancies, the increased IL-6 levels seem a consistent result as they were replicated in a recent study [34,35] (Table 1).

Table 1. Cytokine studies.

Cytokines AND (OCD OR "Obsessive Compulsive Disorder")			
Authors, Date	Subjects	Main Results	Significance
Jiang C. et al. (2018) [36] Meta-analysis	435 cases 1073 controls	TNF-α polymorphisms -> G vs. A model: OR = 1.01; 95% CIs = 0.37–2.77; -> GG vs. AA + AG model: OR = 0.93; 95% CIs = 0.37–2.37; -> GG + AG vs. AA model: OR = 0.22; 95% CIs = 0.06–0.73; -> GG vs. AA model: OR = 0.21; 95% CIs = 0.06–0.71; -> AG + AA model: OR = 0.29; 95% CIs = 0.07–1.16; -> GG + AA vs. AG model: OR = 1.17; 95%CIs = 0.55–2.51;	$p = 0.981$ $p = 0.879$ $p = 0.014$ $p = 0.12$ $p = 0.081$ $p = 0.683$
Colak Sivri R. et al. (2018) [32]	44 OCD patients 40 controls	-> OCD log-TNF-α > controls log-TNF-α -> OCD log-IL-12 < controls log-IL-12 No difference concerning BDNF, TFG-β (tendency of increased level in OCD patients), IL-1β (tendency of decreased level in OCD patients), IL-17, sTNFR1, sTNFR2, CCL3, CCL24 (tendency of increased level in OCD patients), CCL8	$p < 0.001$ $p = 0.014$

Table 1. *Cont.*

Authors, Date	Subjects	Main Results	Significance
Rodriguez N. et al. (2017) [35]	102 OCD patients 47 controls	-> Monocytes percentage of OCD patients > controls -> CD16+ monocytes percentage of OCD patients > controls After LPS stimulation -> OCD-patients IL-1β > controls IL-1β -> OCD-patients IL-6 > controls IL-6 -> OCD-patients GM-CSF > controls GM-CSF -> OCD-patients TNF-α > controls TNF-α -> OCD-patients IL-8 > controls IL-8	$p = 0.005$ $p = 0.004$ $p = 0.049$ $p = 0.041$ $p = 0.013$
Simsek S. et al. (2016) [33]	34 OCD patients 34 controls	-> OCD patients IL-17α > controls IL-17α -> OCD patients TNF-α > controls TNF-α -> OCD patients IL-2 > controls IL-2 No difference for IFNγ, IL-10, IL-6, IL-4 (tendency of increased level in OCD patients)	$p = 0.03$ $p = 0.01$ $p = 0.02$
Rao NP. et al. (2015) [34]	20 OCD patients 20 controls	-> OCD patients IL-2 > controls IL-2 -> OCD patients IL-4 > controls IL-4 -> OCD patients IL-6 > controls IL-6 -> OCD patients IL-10 > controls IL-10 -> OCD patients TNF-α > controls TNF-α No difference concerning IFN-γ	$p = 0.005$ $p = 0.007$ $p = 0.002$ $p = 0.006$ $p = 0.005$
Uguz F. et al. (2014) [37]	7 OCD patients 30 controls	-> cord blood TNF-α of new born infants of women with OCD > cord blood TNF-α of new born infants of control women	$p = 0.036$
Bo Y. et al. (2013) [38]	241 OCD patients 444 controls	IL-1β-511C/T polymorphism No difference between OCD patients and controls	
Zhang X. et al. (2012) [39]	200 OCD patients 294 controls	MCP-1-2518G/A polymorphism No difference between OCD patients and controls	
Liu S. et al. (2012) [40]	187 OCD patients 281 controls	IL-8-251T/A polymorphism No difference	
Gray SM. et al. (2012) [31] Meta-analysis	169 OCD patients 215 controls	-> Decreased IL-1β in OCD patients -> Increased IL-6 in adult free-medication OCD patients No difference concerning IL-6 in OCD children -> Decreased TNF-α in OCD patients without depression No difference in TNF-α when depressed patients are considered	$p < 0.01$ $p = 0.02$ $p < 0.001$
Cappi C. et al. (2012) [22]	183 OCD patients 249 controls	TNF-α A/G polymorphism -> Association of allele A with OCD (χ^2, rs361525)	$p = 0.007$
Fontenelle LF. et al. (2012) [23]	40 OCD patients 40 controls	-> OCD patients CCL3 > controls CCL3 -> OCD patients CXCL8 > controls CXCL8 -> OCD patients sTNFR1 > controls sTNFR1 -> OCD patients sTNFR2 > controls sTNFR2 No difference between OCD and controls concerning CCL2, CCL11, CCL24 (tendency of increased level in OCD patients), CXCL9, CXCL10 (tendency of decreased level in OCD patients), IL-1ra, TNF-α.	$p = 0.03$ $p < 0.001$ $p < 0.001$ $p < 0.01$
Fluitman SB et al. (2010) [24]	10 OCD patients 10 controls	During disgust exposure: -> LPS-stimulated TNF-α in OCD patients decreased after disgust exposure LPS-stimulated TNF-α in controls not changed after disgust exposure -> LPS-stimulated IL-6 in OCD patients decreased after disgust exposure LPS-stimulated IL-6 in control not changed after disgust exposure	$p = 0.07$ $p = 0.040$
Fluitman S. et al. (2010) [25]	26 OCD patients 52 controls	-> OCD patients LPS-stimulated IL-6 < control LPS-stimulated IL-6 No difference concerning LPS-stimulated IL-8 and TNF-α	$p = 0.016$
Hounie AG et al. (2008) [26]	111 OCD patients 250 controls	TNF-α-A/G polymorphism -> Association of the A allele with OCD for 238 G/A and 308 G/A (χ^2)	$p = 0.0005$ and 0.007 respectively

Table 1. *Cont.*

		Cytokines AND (OCD OR "Obsessive Compulsive Disorder")	
Authors, Date	**Subjects**	**Main Results**	**Significance**
Konuk N. et al. (2007) [27]	31 OCD patients 31 controls	-> OCD patients TNF-α > control TNF-α	*p* < 0.001
		-> OCD patients IL-6 > control IL-6	*p* < 0.001
Denys D. et al. (2004) [28]	50 OCD patients 25 controls	-> OCD patients LPS stimulated IL-6 > control LPS stimulated IL-6	*p* = 0.004
		-> OCD patients LPS stimulated TNF-α > control LPS stimulated TNF-α	*p* < 0.001
		-> decreased NK cells activity in OCD patients	*p* = 0.002
		No difference concerning LPS-stimulated IL-10	
Carpenter LL. et al. (2002) [41]	26 OCD patients 26 controls	No difference concerning CSF IL-6 level.	
Monteleone P. et al. (1998) [29]	14 OCD patients 14 controls	-> OCD patients TNF-α < control TNF-α	*p* = 0.001
		No difference concerning IL-6 and IL-1β	
Brambilla F. et al. (1997) [30]	27 OCD patients 27 controls	-> OCD patients IL-1β < control IL-1β	*p* = 0.0004
		-> OCD patients TNF-α < control TNF-α	*p* = 0.0004
Weizman R. et al. (1996) [20]	11 OCD patients 11 controls	No difference concerning IL-1β, IL-2, and IL-3-LA production between OCD patients and controls	
Maes M. et al. (1994) [21]	19 OCD patients 19 controls	No difference concerning IL-1β, IL-6, sIL-2R, sIL-6R	

A/G: adenine/guanine; BDNF: brain-derived neurotrophic factor; C/T: cytosine/thymine; CCL: chemokine ligand; CIs = confidence interval; CSF: cerebrospinal fluid; CXCL: chemokine (C-X-C motif) ligand; G/A: guanine/adenine; GM-CSF: granulocyte-macrophage colony stimulating factor; IFN: interferon; IL: interleukin; IL-1ra: interleukin 1 receptor antagonist; LA: like activity; LPS = lipopolysaccharide; MCP: monocyte chemoattractant protein; NK: natural killer; OCD: obsessive-compulsive disorder; OR = odds ratio; sIL-2R: soluble interleukine-2 receptor; sTNFR: soluble tumor necrosis factor receptor; T/A: thymine/adenine; TNF: tumor necrosis factor.

The fact that IL-6 levels are higher in autoimmune diseases (e.g., rheumatoid arthritis) [19] raises the hypothesis that increased IL-6 levels in OCD could favor the existence of an autoimmune etiological factor. Furthermore, tocilizumab, an IL-6 receptor-neutralizing antibody, appears as a putative treatment in some cases of OCD as this molecule was always found effective in some autoimmune disorders where IL-6 is involved [19]. We could, thus, hypothesize that tocilizumab or other specific immunological treatments could help some specific OCD patients with a possible immunological etiological factor.

It is known also that both IL-6 and TNFα can be involved in asthma pathophysiology and allergic diseases [42], and allergic diseases would appear to be more frequent in OCD patients [43], giving weight to elevated IL-6 and TNFα levels in some OCD cases.

TNFα, IL-1β, and IL-6 are inflammatory cytokines (for a very complete review, see Reference [42]). TNFα is produced by a wide range of cells including T- or B-cells and monocytes (including microglia), and it targets all nucleated cells. TNFα has a complex role, being both pro-inflammatory and immunosuppressive. In the brain, TNFα could be involved in synapses scaling with high levels of TNFα favoring LTP (long term potentiation) and low levels of TNFα favoring LTD (long term depression) [44,45]. Progranulin mutations were found to be associated with hyperexcitability of nucleus accumbens spiny neurons in mice, in line with hyperactivity of cortico-striatal loops in OCD [1], and elevated TNFα levels and hyperactivation of microglia [46]. With the progranulin gene restored, OCD-like behavior disappeared in mice [46]. Frontoparietal dementia patients showing mutations of progranulin presented OCD [46].

IL-1β is also produced by microglia and targets T-cells or endothelial and epithelial cells [42].

IL-6 is produced by both astrocytes and microglia, and IL-6 exposure could increase synaptic activity (for an excellent review on IL-6 central nervous system (CNS) effects, see Reference [47]).

In summary, the literature on cytokines involved in OCD is difficult to interpret as contradictory results are often found. These discrepancies could be due to the heterogeneity of patients studied, e.g., children vs. adults, and they emphasize the importance of considering the immunological status of recruited patients. Thus, differences in symptoms, and their development or response to treatment between OCD patients with and without modified IL-6 and TNFα levels would suggest the possibility of a specific immunological OCD cluster.

3.2. Antibodies

Most studies concerning antibodies in OCD concern pediatric autoimmune neuropsychological disorders associated with streptococcal (PANDAS) infections.

Studies found that a subset of patients suffering from OCD showed high levels of anti-basal ganglia antibodies (ABGAs) and anti-streptolysin O antibodies (ASO) in the blood or corticospinal fluid (CSF) [48–51] (Table 2). These studies strongly support the existence of an autoimmune etiological factor in OCD. However, discrepancies still exist: ABGAs were found in OCD patients (and not in controls), but not in all OCD patients [49] (Table 2).

Table 2. Autoimmunity and OCD.

Antibody AND (OCD OR "Obsessive Compulsive Disorder")			
Authors, Date	**Subjects**	**Main Results**	**Significance**
Akaltun I. et al. (2018) [52]	60 OCD 60 controls	-> Toxoplasma IgG levels related to OCD status -> IgG positivity individuals: increased risk of OCD: OR = 4.84, 95% CIs = 1.78–13.12	$p = 0.001$ $p = 0.002$
Mataix-Cols D. et al. (2017) [53]	30082 OCD 472874 patients	-> Augmentation of the risk to develop autoimmune disease: OR = 1.43; 95% CIs = 1.37–1.49	$p < 0.01$
Flegr J. et al. (2017) [54]	281 men and 831 women not infected 65 men and 350 women infected with toxoplasma	-> Association between toxoplasma infection and OCD: OR = 2.27, 95% CIs = 1.01–5.09	$p = 0.047$
Sutterland AL. et al. (2015) [14] Meta-analysis	No information but 2 studies included	-> Association between OCD status and toxoplasma infection: OR = 3.4; 95% CIs = 1.73–6.68	$p = 0.0004$
Nicolini H. et al. (2015) [55]	37 PANDAS/OCD or tics patients 12 controls	-> OCD patients anti-enolase > controls anti-enolase -> OCD patients anti-streptococcal proteins > controls anti-streptococcal proteins No differences concerning anti-neural antibodies.	$p = 0.035$ $p = 0.05$
Singer HS. et al. (2015) [56]	8 PANDAS/OCD or tics patients 70 controls	No association between clinical exacerbation and anti-tubulin, anti-lysoganglioside GM1, anti D1R, anti D2R titer.	
Frankovich J. et al. (2015) [57]	19 PANS/OCD or eating disorder patients 28 non-PANS but OCD or eating disorder patients	No difference concerning comorbidities (anxiety, mood disorder, irritability, suicidality) No difference concerning Ig levels No difference concerning remitting course, chronic course.	
Cox CJ. et al. (2015) [58]	311 PANDAS/OCD or tics patients 16 controls	-> PANDAS patients anti-D1R patients > controls anti-D1R -> PANDAS patients anti-lysoganglioside > controls anti-lysoganglioside	$p < 0.0001$ $p = 0.0001$
Ebrahimi Taj F. et al. (2015) [59]	76 OCD/ADHD patients 39 controls	-> OCD/ADHD patients anti-streptolysin O > controls anti-streptolysin O -> OCD/ADHD patients anti-streptokinase > controls anti-streptokinase -> OCD/ADHD patients anti-DNase B > controls anti-DNase B	$p < 0.0001$ $p < 0.0001$ $p < 0.0001$
Murphy TK. et al. (2015) [60]	43 PANS/OCD patients	infectious triggers: 58% of GAS, 12% of mycoplasma pneumoniae, 37 of upper respiratory infection, 2% of Lyme No differences between patients with tics and without tics concerning anti-DNase B, ASO, Mycoplasma IgM/IgG, Lyme screen, age of onset, CY-BOCS score, Y-GTSS score	

Table 2. *Cont.*

Authors, Date	Subjects	Main Results	Significance
Murphy TK. et al. (2012) [61]	41 PANDAS/OCD or tic patients 68 non-PANDAS but OCD or tic patients	-> PANDAS patients remissions > non-PANDAS patients remissions	$p < 0.05$
		-> PANDAS patients dramatic onset > non-PANDAS patients dramatic onset	$p < 0.05$
		-> PANDAS patients ASO/anti-DNase > non-PANDAS patients ASO/anti-DNaseB	$p < 0.0001$
		-> remission in PANDAS patients after antibiotic treatment > remission in non-PANDAS after antibiotic treatment	$p < 0.01$
Leckman JF et al. (2011) [62]	31 PANDAS/OCD or tic patients 53 non-PANDAS/OCD or tic patients	No association between clinical exacerbation and new GAS infection.	
Miman O. et al. (2010) [63]	42 OCD patients 100 controls	-> OCD patients anti-toxoplasma IgG > controls anti-toxoplasma IgG	$p < 0.01$
Bhattacharyya S. et al. (2009) [48]	23 OCD patients 23 controls	-> more CSF anti-brain antibody binding to basal ganglia and thalamus for OCD patients than for patients	$p < 0.05$
		-> More CSF glutamate and glycine in OCD patients than in controls	$p < 0.001$
Gause C. et al. (2009) [64]	13 OCD only patients 20 PANDAS/OCD patients 23 PANDAS/tic patients 29 controls	No difference concerning ASO titers	
		No difference concerning serum IgG	
		-> More anti-neural antibodies PANDAS/OCD than in other groups	$p < 0.009$
Morer A. et al. (2008) [49]	32 OCD patients 19 controls	No anti-basal ganglia antibody detected by immunohistochemistry -> Anti-basal ganglia antibodies in OCD patients and no in control detected by immunoscreening No difference concerning ASO titers	
Kirvan CA. et al. (2006) [65]	16 PANDAS/OCD or tic patients 25 non-	-> lysoganglioside GM1 concentration required to inhibit binding PANDAS sera to GlcNAc (an epitope of GAS carbohydrate) < lysoganglioside GM1 concentration required to inhibit binding non-PANDAS sera to GlcNAc (an epitope of GAS carbohydrate)	$p < 0.05$
	PANDAS/OCD or tic or ADHD patients	-> lysoganglioside GM1 = specific inhibitor of PANDAS IgG binding to GlcNAc -> PANDAS sera induced activation of CaM kinase II more than non-PANDAS sera => PANDAS serum responsible for cell signaling	$p = 0.001$
Morer A. et al. (2006) [66]	18 early onset OCD 21 late onset OCD	-> Child OCD ASO titer > adult OCD ASO titer No difference for D8/D17	$p = 0.031$
Singer HS. et al. (2005) [67]	48 PANDAS (OCD or tic status not informed) patients 43 controls	No median ELISA optical density difference concerning serum antibodies No difference concerning reactivity against pyruvate kinase M1, α-enolase, γ-enolase, aldolase C	
Pavone P. et al. (2004) [68]	22 PANDAS (OCD or tic status no informed) patients 22 GAS uncomplicated infected patients	-> PANDAS anti-basal ganglia antibody > GAS patients anti-basal ganglia antibody	$p < 0.001$
		No difference concerning ASO or anti DNase B antibody	
Murphy TK. et al. (2004) [69]	15 OCD or tics patients with large symptom fluctuations 10 OCD or tics patients without large symptom fluctuations	-> positive correlation between OCD severity and ASO titer in patients with large symptom fluctuations	$p = 0.0130$
Luo F. et al. (2004) [70]	47 OCD or tic patients 19 controls	-> OCD or tic patients percentages of D8/D17 positive cells > controls percentages of D8/D17 positive cells	$p = 0.0029$
Inoff-Germain G. et al. (2003) [71]	108 positive children for D8/17 marker 132 negative dor D8/17 marker	No association between D8/17 marker status and OCD or tic status	

Antibody AND (OCD OR "Obsessive Compulsive Disorder")

Table 2. *Cont.*

	Antibody AND (OCD OR "Obsessive Compulsive Disorder")		
Authors, Date	**Subjects**	**Main Results**	**Significance**
Murphy ML. et al. (2002) [72]	12 PANDAS OCD patients	-> abrupt appearance of OCD symptoms -> elevated anti-DNase B titer -> mean age at onset = 7 years	
Eisen JL. et al. (2001) [73]	29 OCD patients 26 controls	No difference in D8/D17 marker positivity	
Murphy TK. et al. (2001) [74]	32 OCD or tic patients 12 controls	-> OCD/tic patients D8/D17 titers > control D8/17 titers	$p = 0.01$
Peterson BS. et al. (2000) [75]	105 tic, OCD or ADHD patients 37 controls	No association between OCD or tic disorder and ASO or anti-DNase B titers -> ASO or anti-DNase B titers positively correlated with putamen or globus pallidus volume in OCD patients	No access to p-values
Marazziti D. et al. (1999) [76]	20 OCD patients 20 controls	-> Increased CD8+ lymphocytes in OCD patients -> decreased CD4+ lymphocytes in OCD patients	$p = 0.002$ $p = 0.003$
Chapman F. et al. (1998) [77]	41 OCD or tic patients 31 controls	-> OCD or tic patients D8/D17 positivity > control D8/D17 positivity	$P < 0.0001$
Khanna S. et al. (1997) [78]	76 OCD patients 55 controls	-> OCD patients mumps and HSV-I IgG > control mumps and HSV-I IgG	$p < 0.05$
Khanna S. et al. (1997) [79]	76 OCD patients 30 controls	-> OCD patients measles CSF IgG < control measles CSF IgG -> OCD patient herpes CSF IgG > control herpes CSF IgG	$p < 0.001$ $p < 0.05$
Murphy TK. et al. (1997) [80]	31 OCD or tic patients 21 controls	-> OCD patients D8/17 positivity > control D8/17 positivity No difference concerning ASO, anti-DNase B and anti-neural antibodies	$p < 0.001$
Swedo SE. et al. (1997) [81]	27 PANDAS/OCD or tic patients 24 controls	-> PANDAS/OCD or tic patients D8/D17 positivity > control D8/D17 positivity	$p < 0.0001$
	Anti-brain antibody and (OCD OR "obsessive–compulsive disorder")		
Bhattacharyya S. et al. (2009) [48]	Cf. antibody AND (OCD OR "obsessive compulsive disorder")		
Dale RC. et al. (2005) [50]	50 OCD patients 40 controls with uncomplicated streptococcal infection	-> ABGA level in OCD patients > ABGA level in controls	$p < 0.005$
Pavone P. et al. (2004) [68]	Cf. antibody AND (OCD OR "obsessive compulsive disorder")		
	ABGA and (OCD OR "obsessive–compulsive disorder")		
Pearlman DM. et al. (2014) [51] Meta-analysis	297 OCD patients 406 controls	-> ABGA seropositivity in OCD patients > ABGA seropositivity in controls	$p < 0.0001$
Dale RC. et al. (2005) [50]	Cf. Anti-brain antibody and (OCD OR "obsessive–compulsive disorder")		

ABGA = anti-basal ganglia antibody; ADHD = attention deficit/hyperactivity disorder; ASO = anti-streptolysin O; CaM Kinase II = Ca^{2+}/calmodulin dependent protein kinase II; CD = cluster of differentiation; CIs = confidence intervals; CSF = cerebrospinal fluid; CY-BOCS = children's Yale–Brown obsessive-compulsive scale; D1R = dopamine 1 receptor;D2R = dopamine 2 receptor; D_ and D17 = B lymphocyte antigen; DNase = deoxyribonuclease; GAS = group A streptococcus; GM1 = monosialotetrahexosylganglioside 1; HSV = herpes simplex virus; Ig = immunoglobulin; OCD: obsessive-compulsive disorder; OR = odds ratio; PANDAS = pediatric autoimmune neuropsychological disorders associated with streptococcal infection; PANS = pediatric acute-onset neuropsychiatric syndrome; Y-GTSS = Yale global tic severity scale.

Furthermore, it was shown that patients suffering from rheumatic fever—a disorder linked with PANDAS—show a higher proportion of a specific B-lymphocyte alloantigen detected with monoclonal antibodies D8/D17 [82,83]. Consequently, some authors tried determining whether the monoclonal antibody D8/D17 could also be used as an OCD marker. The D8/D17 mean value was shown to be higher in OCD or Tourette's child populations vs. control subjects [80], and, as no difference was found between Tourette syndrome and OCD patients in D8/D17 values, it was hypothesized that it could be a marker for OCD in children [80]. Although these results were replicated [74,77], there is still debate surrounding this topic [71,73,81,84] (Table 2).

However, this higher proportion of B-lymphocyte alloantigen detected with monoclonal antibody D8/D17 found in some OCD patients could be a promising way of classifying patients in specific subgroups of OCD (patients with a high D8/D17 value vs. OCD patients without) and could, thus, be a promising line for proposing more specific treatments for these particular patients.

3.3. White Blood Cells

As with cytokines, very contrasting results were found for white blood cells. Some studies found a higher number of CD8+ (cluster od differentiation) lymphocytes and a lower number of CD4+ lymphocytes in OCD patients [76], whereas others did not [85,86] (Table 3). Furthermore, other studies concerning different white blood cells (monocytes or NK cells (natural killer cells)) also found different results [28,35,87] (Table 3). While cytokines are studied intensively in the context of OCD, more studies will need to be done specifically on white-blood-cell counts and activity.

Table 3. White blood cells and OCD.

Authors, Date	Subjects	Main Results	Significance
"White Blood Cells" OR "Total Blood Count" AND (OCD OR "Obsessive Compulsive Disorder")			
Atmaca M. et al. (2011) [85]	30 OCD patients 30 controls	-> OCD patients neutrophils < control neutrophils	$p < 0.05$
lymphocyte AND ("obsessive compulsive disorder" OR OCD)			
Marazziti D. et al. (2009) [88]	18 OCD patients	-> CD8+ lymphocytes cells decreased after treatment	$p = 0.004$
		-> CD4+ lymphocytes increased after treatment	$p = 0.005$
Denys D. et al. (2006) [87]	42 OCD patients	No effect of paroxetine or venlafaxine on TNF-α, IL-4, IL-6, IL-10, IFN-γ, NK cell activity, monocytes, T-cells, and B-cells percentages	
Denys D. et al. (2004) [28]		Cf. cytokines AND (OCD OR "obsessive compulsive disorder")	
Marazziti D. et al. (2003) [89]	10 OCD patients 10 controls	-> OCD patients (3)H-paroxetine-binding density < controls (3)H-paroxetine-binding density	$p = 0.0001$
Eisen JL. et al. (2001) [73]		Cf. antibody AND (OCD OR "obsessive compulsive disorder")	
Murphy TK. et al. (2001) [74]		Cf. antibody AND (OCD OR "obsessive compulsive disorder")	
Marazziti D. et al. (2001) [90]	10 OCD patients 15 controls	Presence of 5-HT2C and 5-HT2A mRNAs in patients and controls	
Rocca P. et al. (2000) [91]	15 OCD patients 10 controls	-> decrease of peripheral benzodiazepine receptor mRNA	$p < 0.05$
Marazziti D. et al. (1999) [76]		Cf. antibody AND (OCD OR "obsessive compulsive disorder")	
Ravindran AV. et al. (1999) [92]	26 OCD patients 16 controls	-> OCD patients circulating NK cells < control circulating NK cell	$p < 0.05$
		No difference concerning B or T cells	
		No difference in circulating NK cells after treatment.	
Chapman F. et al. (1998) [77]		Cf. antibody AND (OCD OR "obsessive compulsive disorder")	
Murphy TK. et al. (1997) [80]		Cf. antibody AND (OCD OR "obsessive compulsive disorder")	
Swedo SE. et al. (1997) [81]		Cf. antibody AND (OCD OR "obsessive compulsive disorder")	

Table 3. *Cont.*

"White Blood Cells" OR "Total Blood Count" AND (OCD OR "Obsessive Compulsive Disorder")			
Authors, Date	Subjects	Main Results	Significance
Barber Y et al. (1996) [86]	7 OCD patients 9 controls	No difference in lymphocytes between OCD patients and OCD	
		No difference in lymphocytes after treatment.	
Rocca P. et al. (1991) [93]	18 OCD patients 50 controls	-> Number of binding sites for peripheral benzodiazepine receptor lower in OCD patients	*p* < 0.05
monocytes AND ("obsessive compulsive disorder" OR OCD)			
Rodriguez N et al. (2017) [35]		Cf. cytokines AND (OCD OR "obsessive compulsive disorder")	
Denys D. et al. (2006) [87]		Cf. lymphocyte AND ("obsessive compulsive disorder" OR OCD)	
Denys D. et al. (2004) [28]		Cf. lymphocyte AND ("obsessive compulsive disorder" OR OCD)	
Weizman R. et al. (1996) [20]		Cf. cytokines AND (OCD OR "obsessive compulsive disorder")	
NK cells" AND ("obsessive compulsive disorder" OR OCD)			
Denys D. et al. (2004) [28]		Cf. lymphocyte AND ("obsessive compulsive disorder" OR OCD)	
Ravindran V. et al. (1999) [92]		Cf. lymphocyte AND ("obsessive compulsive disorder" OR OCD)	

CD = cluster of differentiation; OCD = obsessive-compulsive disorder; HT2A = serotonin 2A; HT2C = serotonin 2C; IFN = interferon; mRNA = messenger ribonucleic acid; NK = natural killer; TNF = tumor necrosis factor.

4. Infections and OCD

Here, we describe two of these infectious etiological factors (Table 4), and we discuss some possible mechanisms via which these infectious agents could lead to OCD, and hence, why these infection contexts could be considered as specific OCD subtypes.

Table 4. Infectious agents and OCD.

Infection AND (OCD OR "Obsessive Compulsive Disorder")			
Authors, Date	Subjects	Mains Results	Significance
Ursoiu F. et al. (2018) [94]	101 HIV patients	No association between HIV and OCD	
Akaltun I. et al. (2018) [52]		Cf. antibody AND (OCD OR "obsessive compulsive disorder")	
Flegr J et al. (2017) [54]		Cf. antibody AND (OCD OR "obsessive compulsive disorder")	
Sutterland AL. et al. (2015) [14] Meta-analysis		Cf. antibody AND (OCD OR "obsessive compulsive disorder")	
Nicolini H. et al. (2015) [55]		Cf. antibody AND (OCD OR "obsessive compulsive disorder")	
Miman O. et al. (2010) [63]		Cf. antibody AND (OCD OR "obsessive compulsive disorder")	
Dale RC. et al. (2004) [95]	40 patients with post-streptococcal dyskinesias	-> 27.5% of these patients suffered from OCD	

Table 4. *Cont.*

Authors, Date	Subjects	Mains Results	Significance
Infection AND (OCD OR "Obsessive Compulsive Disorder")			
Giulino L. et al. (2002) [96]	83 OCD patients	-> OCD patients with upper respiratory infection had more sudden onset than patients without upper respiratory infection No difference concerning tic or ADHD comorbidity between OCD patients with or without upper respiratory infection.	$p = 0.02$
Lougee L. et al. (2000) [97]	54 PANDAS/OCD or tic patients 139 first relatives	-> 26% of OCD patients had a relative suffering from OCD	
lyme AND ("obsessive compulsive disorder" OR OCD)			
Johnco C. et al. (2018) [98]	147 patients with Lyme disease	-> 84% of patients reported obsessive compulsive symptoms -> 90.9% of patients reported gradual onset of symptoms -> 47% of patients were treated with psychotropic treatment and 76.9% of them presented at least partial improvement -> 50.9% of patients treated with antibiotics reported at least partial improvement in symptoms	
Streptococcus AND (OCD OR "obsessive compulsive disorder")			
Stagi S. et al. (2018) [99]	179 PANDAS/OCD or tic patients	-> reduced vitamin D in PANDAS patients	$p < 0.0001$
Mataix-Cols D. et al. (2017) [53]	Cf. antibody AND (OCD OR "obsessive compulsive disorder")		
Calaprice D. et al. (2017) [100]	698 PANS patients	-> age of onset between 7 and 8 years -> 88% of sudden onset -> 87% of patients presented recurrences -> 94% of patients presented a history of OCD -> 71% with motor tics and 57% with vocal tics	
Wang HC. et al. (2016) [101]	2596 patients infected with GAS 25960 controls	-> increased risk of tic disorder in GAS infected patients	No full access
Nicolini H. et al. (2015) [55]	Cf. antibody AND (OCD OR "obsessive compulsive disorder")		
Frankovich J. et al. (2015) [57]	Cf. antibody AND (OCD OR "obsessive compulsive disorder")		
Ebrahimi Taj F. et al. (2015) [59]	Cf. antibody AND (OCD OR "obsessive compulsive disorder")		
Murphy TK. et al. (2012) [61]	Cf. antibody AND (OCD OR "obsessive compulsive disorder")		
Leckman JF. et al. (2011) [62]	Cf. antibody AND (OCD OR "obsessive compulsive disorder")		
Murphy TK. et al. (2010) [102]	107 OCD or tic patients	-> 17.8% of patients had mother suffering from autoimmune disease	
Gause C. et al. (2009) [64]	Cf. antibody AND (OCD OR "obsessive compulsive disorder")		
Kurlan R. et al. (2008) [103]	40 PANDAS/OCD or tic patients 40 non-PANDAS/OCD or tics	No difference in the number of exacerbations (but a strong tendency for increased exacerbation risk, $p = 0.07$). -> more frequent GAS infection associated with exacerbation	$p = 0.002$
Dale RC. et al. (2004) [95]	Cf. infection AND (OCD OR "obsessive compulsive disorder")		
Luo F. et al. (2004) [70]	Cf. antibody AND (OCD OR "obsessive compulsive disorder")		

Table 4. *Cont.*

Infection AND (OCD OR "Obsessive Compulsive Disorder")			
Authors, Date	**Subjects**	**Mains Results**	**Significance**
Pavone P. et al. (2004) [68]	Cf. antibody AND (OCD OR "obsessive compulsive disorder")		
Murphy TK. et al. (2004) [69]	Cf. antibody AND (OCD OR "obsessive compulsive disorder")		
Giulino L. et al. (2002) [96]	Cf. infection AND (OCD OR "obsessive compulsive disorder")		
Murphy TK. et al. (2001) [74]	Cf. antibody AND (OCD OR "obsessive compulsive disorder")		
Lougee L. et al. (2000) [97]	Cf. infection AND (OCD OR "obsessive compulsive disorder")		
Giedd JN. et al. (2000) [104]	34 PANDAS/OCD or tics 82 controls	-> PANDAS patients mean caudate volume > controls mean caudate volume -> PANDAS patients mean putamen volume > controls mean putamen volume -> PANDAS patients mean globus pallidus volume > controls mean globus pallidus volume No difference for thalamus and total brain volume	$p = 0.004$ $p = 0.02$ $p = 0.02$
Swedo SE. et al. (1998) [105]	50 PANDAS patients	-> Mean age at onset: 7.4 years -> tics and OCD: 64%; tics only: 16% and OCD only: 20% -> ADHD comorbidity: 40%, ODD comorbidity: 40%, MDD comorbidity: 36%	
Murphy TK. et al. (1997) [80]	Cf. antibody AND (OCD OR "obsessive compulsive disorder")		
toxoplasma (OCD OR "obsessive compulsive disorder")			
Akaltun I. et al. (2018) [52]	Cf. antibody AND (OCD OR "obsessive compulsive disorder")		
Flegr J et al. (2017) [54]	Cf. antibody AND (OCD OR "obsessive compulsive disorder")		
Sutterland AL. et al. (2015) [14]	Cf. antibody AND (OCD OR "obsessive compulsive disorder")		
Miman O. et al. (2010) [63]	Cf. antibody AND (OCD OR "obsessive compulsive disorder")		

ADHD = attention deficit/hyperactivity disorder; GAS = Group A streptococcus; HIV = human immunodeficiency virus; MDD = major depressive disorder; ODD = oppositional defiant disorder; PANDAS = pediatric autoimmune neuropsychological disorders associated with streptococcal infection; PANS = pediatric acute-onset neuropsychiatric syndrome.

4.1. Streptococcal Infection

Streptococcus pyogenes is a bacterial group that can lead to several pathologies such as pharyngitis, scarlet fever, or erysipelas [106]. Among these diseases, rheumatic fever is the one we were interested in. This disease is characterized by elevated ASO (anti-streptolysin O) or anti-DNAse B antibody levels [107], and it affects the heart, skin, bone joints, and CNS [108,109]. The Jones criteria are usually used to make the diagnosis. They consist of carditis, arthritis, chorea, erythema marginatum, and subcutaneous nodules with evidence of *S. pyogenes* infection [108,109]. *S. pyogenes* can, thus, affect the nervous system through choreic movements. This chorea, called Sydenham's chorea [110,111], is characterized by involuntary movements which are irregular, rapid, and transient, and which are typically manifested in the extremities [111,112]. Sydenham's chorea is characterized by antibodies found in the basal ganglia that react with *N*-acetyl-beta-D-glucosamine of *S. pyogenes*, and with lysoganglioside and tubulin of the brain [111,113]. This cross-reaction is made possible by a mimicry process [111]. Furthermore, it was shown recently by Cox and colleagues that these antibodies could react with the D2-receptor (D2R) complex, which could be causal in Sydenham's chorea, as risperidone reverses this movement disorder [114]. In summary, antibodies that originally target *S. pyogenes* may also attack the patient's brain.

Since the basal ganglia (where Sydenham's chorea antibodies are found) appear to be a key region in OCD neurobiology [1,3], one could imagine that antibodies against basal ganglia (which seems the case in PANDAS [68]) could impair their functioning and lead to OCD symptoms in some conditions. In this context, it is notable that OCD may be associated with Sydenham's chorea [115]. In addition to this association, the concept of PANDAS (pediatric autoimmune neuropsychological disorder associated with streptococcal infection) was originally defined by Swedo et al. in 1998 [105] as follows: presence of OCD or tic disorder, symptom onset between the age of three and puberty, exacerbation of symptoms associated with streptococcal infection, and presence of neurological abnormalities during symptom exacerbation, but in the absence of frank chorea which would suggest Sydenham's chorea [105]. This original description of PANDAS was modified in 2012 by Swedo et al. to become PANS (pediatric acute-onset neuropsychiatric syndrome, with abrupt onset of OCD or severely restricted food intake and presence of additional neuropsychiatric symptoms such as anxiety, emotional liability, etc.) [13]. PANDAS and PANS could, thus, constitute a specific OCD subgroup for which the underlying physiological mechanism could be the same, that is to say, an autoantibody against basal ganglia neurons [110,113].

However, no D2R antibodies were found in PANDAS patients [12]. Cox et al. recently studied patients with tic disorders or OCD or both, and with a history of streptococcal infection [58]. They found that these patients as a whole presented elevated levels of anti D1-receptor (D1R) antibodies in the serum (with elevated anti-lysoganglioside antibodies) compared to controls. As in Sydenham's chorea, anti-lysoganglioside antibodies seem to be involved [58,65] (Table 4).

Recently, antibodies in children suffering from PANDAS were found to bind more to cholinergic interneurons of mice than control antibodies when mice were infused with patient and control serum in their striatum [116]. Taken together, these results raise the question of the proportions of dopamine receptor subtypes and the role of cholinergic interneurons in OCD and more particularly in PANDAS, which is a good example of the multiple etiologies of OCD. It is one of the rare clearly identified etiological factors of OCD. About 5% of pediatric OCD patients meet the criteria for PANDAS (or PANS) [117] and it is important to distinguish this etiology from others in OCD patients. Indeed, the prognosis of PANDAS seems relatively good, as Leon et al. found that 88% of children originally suffering from PANDAS with moderate-to-severe OCD presented no OCD symptoms (55%) or only subclinical symptoms (33%) after approximately three years of follow-up [118] (Table 5). This result of a good prognosis is confirmed by Murphy et al. [61], but not by Frankovich et al. [57] (Table 5). By contrast, 48% of OCD patients were found to be still symptomatic after 30 years [119]. However, as this study began in 1954, and PANDAS patients, which represent about 5% of OCD patients, were first described in 1998 [105,117], one may hypothesize that PANDAS and non-PANDAS OCD patients were pooled together. Furthermore, treatment of PANDAS (described below) is not identical to OCD treatment. Prophylactic antibiotics or antibiotic treatment, anti-inflammatory treatment, and immunoglobulin, indicated in PANDAS treatment, are not prescribed in "idiopathic OCD" [118,120–125]. Therefore, it would be important to recognize and adequately treat PANDAS within a personalized medical setting.

Table 5. Specific immunological treatment in OCD.

(PANDAS OR PANS) AND Treatment AND (OCD OR "Obsessive Compulsive Disorder")			
Authors, Date	Subjects	Main Results	Significance
Leon J. et al. (2018) [118]	33 PANDAS patients	Follow-up lasted between 2.2 and 4.8 years	
		Initially, all patients treated with antibiotics	
		During the follow-up period, 45% of patients took psychotropic treatments	
		At the time of follow-up, 18 patients presented no symptoms, 11 only subclinical symptoms, 3 moderate symptoms, and 1 severe symptom.	
Calaprice D. et al. (2018) [120]	698 PANS patients	675 patients treated with antibiotics, 437 with anti-inflammatories, 378 with psychotropic treatments	
		52% of "very effective" treatments with antibiotics	
		NSAIDs were at least "somewhat effective" for 80% of patients	
		Steroids were at least "somewhat effective" for 72% of patients	
		IVIG were at least "somewhat effective" for 74% of patients	
Brown K. et al. (2017) [121]	98 PANS patients	-> duration of symptomatic periods treated with steroids < duration of symptomatic periods of non-treated patients	$p < 0.001$
		-> shorter symptomatic periods when initially treated with steroids	$p < 0.01$
Brown KD. et al. (2017) [126]	95 PANS patients	-> Symptomatic periods treated with NSAID lasted shorter than non-treated symptomatic periods	$p < 0.0001$
		-> the more the duration without treatment is short, the more symptomatic period were short	$p = 0.02$
Spartz EJ. et al. (2017) [122]	159 PANS patients	No clinical data allow to distinguish responders and non-responders to NSAIDs	
		31% of patients with NSAID increases reported improvement in symptoms	
		35% of patients with NSAID removal reported symptom increases after removal	
Murphy TK. et al. (2017) [123]	31 PANS patients (17 with azithromycin, 14 with placebo)	-> azithromycin group improvement > non-azithromycin group (CGI)	$p = 0.003$
		No difference on CY-BOCS	
Calaprice D. et al. (2017) [100]	Cf. Streptococcus AND (OCD OR "obsessive compulsive disorder")		
Williams KA. et al. (2016) [124]	35 PANDAS patients (IVIG group = 17, placebo group = 18)	-> At week 6 (double blind phase): no difference between groups (CY-BOCS)	$p < 0.0001$
		-> Improvement after open label IVIG (CY-BOCS)	
Nadeau JM. et al. (2015) [127]	11 PANS patients partially responder to antibiotics	-> Improvement after CBT (CY-BOCS)	$p = 0.01$
Nicolini H. et al. (2015) [55]	Cf. antibody AND (OCD OR "obsessive compulsive disorder")		
Frankovich J. et al. (2015) [57]	Cf. antibody AND (OCD OR "obsessive compulsive disorder")		
Latimer ME. et al. (2015) [128]	35 PANDAS patients	-> 6 months after therapeutic plasma apheresis: improvement of 65% (local questionnaire)	

Table 5. *Cont.*

(PANDAS OR PANS) AND Treatment AND (OCD OR "Obsessive Compulsive Disorder")			
Authors, Date	Subjects	Main Results	Significance
Demesh D. et al. (2015) [129]	10 PANDAS patients	-> Improvement in symptom intensity after antibiotic treatment (local questionnaire)	$p = 0.03$
		-> Improvement in symptom intensity after tonsillectomy (local questionnaire)	$p = 0.02$
Ebrahimi Taj F. et al. (2015) [59]	Cf. antibody AND (OCD OR "obsessive compulsive disorder")		
Pavone P. et al. (2014) [130]	120 PANDAS patients (56 patients with tonsillectomy or adrenotonsillectomy, 64 without)	No difference concerning symptomatology, streptococcal antibodies or anti-neural antibodies (evaluation every two months for 2 years)	
Murphy TK. et al. (2012) [61]	Cf. antibody AND (OCD OR "obsessive compulsive disorder")		
Bernstein GA. et al. (2010) [131]	21 PANDAS patients 18 non-PANDAS OCD patients	No difference concerning age at onset of OCD	
		No difference concerning CY-BOCS score	
		-> PANDAS patients YGTSS score > non-PANDAS patients YGTSS score	$p = 0.013$
		No difference concerning ASO or anti-DNase B titers	
		-> In non-PANDAS OCD patients, separation anxiety disorder and social phobia are more frequent	$p = 0.02$ and 0.047 respectively
Storch EA. et al. (2006) [132]	7 PANDAS patients	-> CY-BOCS improvement after 3 weeks of CBT	$p = 0.018$
Snider LA. et al. (2005) [133]	23 PANDAS patients	-> Decrease in number of symptom exacerbations with antibiotic treatment	$p < 0.01$
Garvey MA. et al. (1999) [134]	37 PANDAS patients (double blind and cross over design)	No difference in symptoms following antibiotic treatment	
Swedo SE. et al. (1998) [105]	Cf. Streptococcus AND (OCD OR "obsessive compulsive disorder")		
NSAID and (OCD OR "obsessive–compulsive disorder")			
Brown KD. et al. (2017) [126]	Cf. (PANDAS OR PANS) AND treatment AND (OCD OR "obsessive compulsive disorder")		
Spartz EJ. et al. (2017) [122]	Cf. (PANDAS OR PANS) AND treatment AND (OCD OR "obsessive compulsive disorder")		
Shalbafan M. et al. (2015) [135]	25 OCD patients with celecoxib (+SRI) 25 OCD patients with placebo (+SRI)	-> lower CY-BOCS score at week 10 in celecoxib group than in placebo group	$p = 0.047$
Sayyah M. et al. (2011) [136]	27 OCD patients with celecoxib (+fluoxetine) 25 OCD patients with placebo (+fluoxetine)	-> lower CY-BOCS score at week 8 in celecoxib group than in placebo group	$p = 0.037$
		-> significant effect of time-by-treatment interaction in ANOVA	$p = 0.018$
"anti-inflammatory" and (OCD OR "obsessive–compulsive disorder")			
Calaprice D. et al. (2018) [120]	Cf. (PANDAS OR PANS) AND treatment AND (OCD OR "obsessive compulsive disorder")		
Brown K. et al. (2017) [121]	Cf. (PANDAS OR PANS) AND treatment AND (OCD OR "obsessive compulsive disorder")		
Brown KD. et al. (2017) [126]	Cf. (PANDAS OR PANS) AND treatment AND (OCD OR "obsessive compulsive disorder")		

Table 5. *Cont.*

(PANDAS OR PANS) AND Treatment AND (OCD OR "Obsessive Compulsive Disorder")			
Authors, Date	**Subjects**	**Main Results**	**Significance**
Shalbafan M. et al. (2015) [135]	Cf. NSAID and (OCD OR "obsessive–compulsive disorder")		
Sayyah M. et al. (2011) [136]	Cf. NSAID and (OCD OR "obsessive–compulsive disorder")		
minocycline and (OCD OR "obsessive–compulsive disorder")			
Esalatmanesh et al. (2016) [137]	47 OCD patients with minocycline (+fluvoxamine) 47 OCD patients with placebo (+fluvoxamine)	-> lower Y-BOCS score at week 10 in minocylcine group than in placebo group	$p = 0.008$
Rodriguez CI. et al. (2010) [138]	9 OCD patients with minocycline (+SRI)	No effect of minocycline at week 12	
N-acetylcysteine and (OCD OR "obsessive–compulsive disorder")			
Ghanizadeh A. et al. (2017) [139]	18 OCD patients with NAC (+citalopram) 11 OCD patients with placebo (+citalopram)	-> lower Y-BOCS score at week 12 in NAC group than in placebo group	$p < 0.02$
Costa DLC. et al. (2017) [140]	40 OCD patients randomized in 2 groups: NAC and placebo (no access to the details)	-> No difference between the two groups concerning Y-BOCS scores.	
Paydary K. et al. (2016) [141]	23 OCD patients with NAC (+fluvoxamine) 23 OCD patients with placebo (+fluvoxamine)	-> No difference between the two groups concerning Y-BOCS at week 10.	
Sarris J. et al. (2015) [142]	22 OCD patients with NAC (+TAU) 22 OCD patients with placebo (+TAU)	-> No difference between the two groups concerning Y-BOCS at week 16.	
Afshar F. et al. (2012) [143]	24 OCD patients with NAC (+SRI) 24 OCD patients with placebo (+SRI)	-> lower Y-BOCS score at week 12 in NAC group than in placebo group	$p = 0.03$

ASO = anti-streptolysin O; CBT = cognitive behavioral therapy; CGI = clinical global impression; CY-BOCS = Children's Yale–Brown Obsessive Compulsive Scale; IVIG = intravenous immunoglobulin; NAC = N-acetylcysteine; NSAID = non-steroidal anti-inflammatory drug; PANDAS = pediatric autoimmune neuropsychological disorders associated with streptococcal infection; PANS = pediatric acute-onset neuropsychiatric Syndrome; SRI = serotonin reuptake inhibitor; TAU = treatment as usual; Y-BOCS = Yale-Brown Obsessive Compulsive Scale; Y-GTSS = Yale Global Tic Severity Scale.

4.2. Toxoplasma gondii

Toxoplasma gondii is an intracellular parasite that is linked to several psychiatric disorders including schizophrenia and bipolar disorder [144,145]. *T. gondii* is also linked to OCD [14,54,63,145] (Table 4). A recent study found that the presence of anti-*Toxoplasma gondii* IgG (immunoglobulin G) in serum was more frequent in OCD patients than in controls (the odds ratio (OR) was 4.84 (confidence intervals = 1.78–13.12) in favor of OCD) [52] (Table 4). Furthermore, in a 1991 study, Strittmatter and colleagues showed that the CNS areas most affected by *T. gondii* were the cerebral hemispheres (91%) and the basal ganglia (78%) which are implicated in OCD neurobiology [146]. There are several hypotheses regarding how *T. gondii* reaches the CNS (for a review, see the article by Ueno et al. [147]). Among these, the monocyte hypothesis is of particular interest. Indeed, the fact that *T. gondii* is found in the brain CD11b+ monocytes, which can be microglial cells (the resident monocytes of the brain) [148], suggests that *T. gondii* can invade monocytes in the peripheral blood supply and then reach the brain.

Once in the brain and in microglia, these monocytes become activated and show increased migratory activity [149].

T. gondii infection leads to IFN-γ (interferon) production, and then, to the induction of IDO (indolamine-2,3-dioxygenase), mainly produced by microglia and one of the main enzymes of kynurenine pathway [150–154]. This induction of IDO by *T. gondii* occurs firstly in parallel with the *T. gondii*-induced microglia activation and can secondly lead to a tryptophan depletion (since the kynurenine pathway is part of tryptophan catabolism) [149,150,155]. As tryptophan is the essential amino acid for serotonin synthesis, tryptophan depletion could interfere with OCD physiological pathways since OCD symptoms are improved with specific serotonin reuptake inhibitors (SSRIs) [156–158]. Nonetheless, we have to keep in mind that this causal tryptophan depletion hypothesis is still a matter of debate in MDD (major depressive disorder); thus, the putative *T. gondii* role in OCD is unclear [159].

We could also hypothesize a *T. gondii* action at the level of striatal dopamine receptors. *T. gondii* contains genes coding for tyrosine hydroxylase, and it was shown that *T. gondii* increases the dopamine release [160]; therefore, *T. gondii* could lead to OCD through dopamine release and its action on striatal D1 receptors, and then, via the activation of the direct pathway (associated with the D1 receptor [161]). Nonetheless, this hypothesis is highly speculative since interactions among D1 receptors, D2 receptors (between direct and indirect pathway), and serotonin receptors are complex, and D1 receptor downregulation could be a consequence of a D1 receptor hyper stimulation, thus leading to an inhibition of the direct pathway.

Finally, there is a neurotoxic hypothesis, via the direct neurotoxic role of quinolinic acid (produced by the kynurenine pathway) and IFN-γ which acts as a neurotoxic agent through its action on the kynurenine pathway [162]. Therefore, one could speculate that *T. gondii* is neurotoxic for the striatal microcircuit, and thus, contributes to the occurrence of OCD symptoms.

According to these different hypotheses on the role of *T. gondii* in the genesis of OCD, some innovative treatment options might be suggested such as the use of IDO inhibitors used for some cancer treatments [163], which were already tested in some animal models of schizophrenia where such treatment seems to protect the striatum from the negative effects of kynurenine pathway activation [164].

5. Alternative Treatments for OCD

The above distinct etiological factors in OCD could be taken into account to develop specific treatments. Tricyclic or SSRI antidepressants are the usual treatment for OCD [165]. For refractory and severe OCD, deep brain stimulation can also be used [6]. Other treatments were also developed for specific etiological factors.

5.1. Specific Treatment in the PANS/PANDAS Context

Several specific treatments were studied for PANS/PANDAS patients. Intravenous immunoglobulin (IVIG) could be an effective treatment [120,124,166] (Table 5); however, its effectiveness remains to be confirmed. Another treatment procedure studied was apheresis. Two studies found this treatment to be effective [128,166] (Table 5), but they suffered from limitations (absence of a control group or a limited number of patients studied), which meant no definitive conclusion could be drawn on its effectiveness in PANDAS. The effects of antibiotic treatment in PANDAS were also studied. Four studies without control groups found that antibiotics could be effective [118,120,129,133] (Table 5), although a study comparing azithromycin vs. placebo as a treatment for PANS over four weeks failed to find an effect of azithromycin on OCD symptoms as measured with the Children's Yale–Brown Obsessive Compulsive Scale (CY-BOCS) [123] (Table 5). However, if streptococcal infection is still present during acute episodes of PANDAS, antibiotics are considered as the best treatment [167,168]. Finally, corticosteroid and nonsteroidal anti-inflammatory drugs (NSAIDs) do appear to be effective [120–122,126] (Table 5).

Hence, several alternative specific treatments to PANDAS/PANS were studied. However, even if some of these proposed treatments seem promising, robust clinical evidence is still lacking to allow us to reach a definitive conclusion [169].

5.2. Specific Treatment in the "Classical" OCD Context

In addition to PANS/PANDAS, immunological treatments were also tested in "classical" OCD, that is to say, with no clearly identified etiological factor. NSAIDs show contrasting results [135,136,170] (Table 5). However, in the general context of OCD with no specific etiological factors, anti-inflammatory treatment seems to have a place in the treatment strategy, which could be more precisely defined if OCD etiologies were better known. Minocycline, a specific antibiotic, is particularly interesting because of its action on microglia (see below). Minocycline was studied as a potential new pharmacological treatment for OCD, and the results were mixed: one study found that minocycline could be a good adjunctive treatment to classical OCD treatment with SSRIs [137], but another one did not find this result [138] (Table 5). Another antibiotic, cefdinir, was studied, but it showed no effectiveness on the CY-BOCS scores when compared to a placebo [171]. Finally, N-acetylcysteine (NAC), an antioxidant which has a neuroprotective role against oxidative stress, produced divergent results [139–143] (Table 5).

These different studies on immunological treatment in the PANDAS/PANS contexts or otherwise indicate that some specific treatments for different aspects of immunity could have a place in OCD treatment.

6. Conclusions: Future Lines of Research for Etiological Immune Response Factors

6.1. Animal Models

The rodent animal model is widely used in anxiety disorder studies. Rodents present many behavioral signs of anxiety in various contexts. Among these behaviors, grooming was considered by some authors as a compulsive-like behavior, due to its repetitive and sequential organization [172]. Hence, several animal models that were created by mutating genes of interest (e.g., *Sapap3* mutant-mice [173]) were considered as animal models for OCD because of their excessive grooming behavior among other parameters. *Hox* genes are homeotic genes [174], which are responsible for the anterior posterior segmentation of the organism. They are also involved in the formation of the hematopoietic system, and *Hoxb8* is especially involved in the differentiation of myeloid progenitor cells, one source of microglial cells [175]. It is, therefore, of note that, firstly, *Hoxb8* mutant-mice show exacerbated grooming behavior since, in the brain, microglia are the only cells linked to *Hoxb8*, and secondly, that grooming could be reversed after normal bone marrow transplantation which allows *Hoxb8*-derived microglia to migrate to the brain [175]. These data show a direct involvement of microglia, which is an immunological component in compulsive-like behaviors. Therefore, microglia (see below) could be a promising future line of research to better understand OCD and the role of immunology in a specific OCD cluster.

Rats exposed to *Streptococcus* antigens show more grooming behavior than control rats [176] and offer a model within which to investigate the link between OCD neurobiology and PANDAS. Indeed, grooming behavior was found to be reversible with serotonin re-uptake inhibitors, and IgG was found in key brain areas of OCD rats (i.e., striatum, thalamus, and frontal cortex), and glutamate and dopamine levels were also found to be modified [176]. Furthermore, the amelioration of some of the previous abnormalities found in this model with an antibiotic treatment [177] is in line with other results in humans [123].

6.2. Microglia

Tourette syndrome is a condition close to the OCD spectrum [178,179], and is well known for showing interneuron loss [180]. Interestingly, a high level of CD45, which is a marker of activated microglia, was found in Tourette's post-mortem basal ganglia [181,182]. An elevated expression of CCL2 (chemokine ligand 2), which is a chemokine that activates microglia, was also found in these brains [181,182], raising the question of whether the interneuron loss in Tourette syndrome is linked in some manner to microglia activation. Indeed, among the functions of microglia are synapse elimination and phagocytosis [183]. There is convergent evidence in OCD, as a recent study found microglia activation in an OCD brain employing a PET (positron emission tomography) protocol [184], which shows a potential role for inflammation and microglia in OCD neurobiology [11]. These results are consistent with the cytokine levels found in OCD patients, as activated microglia produce IL-6, IL-1β, and TNF-α [11,181]. This could explain the effectiveness of minocycline, which reduces microglia activation [185], as another OCD treatment in Reference [137], and highlights the importance for precision medicine to consider immunological etiological factors.

6.3. The Importance of the Attempt to Identify Different OCD Etiologies

As we can see (Table 6), it is likely that there are multiple etiological factors in OCD. Genetic and environmental factors clearly play a role in the emergence of OCD. Some genetic studies indicate the involvement of immune response genes in the physiological basis of OCD [9,10,186]. The environment could play a role through epigenetic processes [187], and could also have a more direct influence on brain function through immunological processes, as is the case with PANDAS. Furthermore, it was shown that stress may directly impact some immunological parameters [188], raising the putative role of psychological stressors through immunological responses in OCD emergence.

Table 6. Summary.

Divergent results concerning cytokines (especially IL-6, TNF-α) were found between studies. These discrepancies, therefore, raise the question of different patient populations, with some patients possibly presenting with immunological deficiencies, thus explaining the discrepancies.
Antibody studies show that autoimmune factors could be specific etiologies in OCD.
Streptococcus pyogenes is already recognized as possibly leading to OCD through PANS (pediatric acute-onset neuropsychiatric syndrome), as is *Toxoplasma gondii*. The mechanisms leading to OCD for *S. pyognes* and *T. gondii* are still unknown, but autoimmunity seems to be involved.
According to these different possible immune etiological factors (autoimmunity, infection), some specific treatments were already tested opening the way to individualized specific treatments. An effort to clearly distinguish between the different etiological (including immunological) factors is still necessary in order to develop more effective OCD treatments

IL = interleukin; OCD = obsessive-compulsive disorder; PANS = pediatric acute-onset neuropsychiatric syndrome; TNF = tumor necrosis factor.

Future research should focus on these etiological factors (genetic, immunological, etc.) in order to elucidate the biological bases of OCD, and to develop prevention tools and better treatments [189], paving the way to precision individualized therapies [190] for the benefit of patients. The identification of more specific biological clusters in OCD is essential in order to advance our knowledge and treatment of OCD.

Author Contributions: H.L. and L.M. conceived and designed the research on Pubmed database; H.L. performed the research on Pubmed database, analyzed the results and wrote the paper; L.M. rearranged the text and improved the text; P.S. and A.P contributed rearrange and contributed to improvements with commentaries; J.-M.B. contributed to improvements with commentaries.

Funding: This research received no external funding.

Conflicts of Interest: The authors declare no conflicts of interest.

References

1. Pauls, D.L.; Abramovitch, A.; Rauch, S.L.; Geller, D.A. Obsessive-compulsive disorder: An integrative genetic and neurobiological perspective. *Nat. Rev. Neurosci.* **2014**, *15*, 410–424. [CrossRef] [PubMed]

2. American Psychiatric Association; DSM-5 Task Force. *Diagnostic and Statistical Manual of Mental Disorders: DSM-5*, 5th ed.; American Psychiatric Association: Washington, DC, USA, 2013; 947p.

3. Abramowitz, J.S.; Taylor, S.; McKay, D. Obsessive-compulsive disorder. *Lancet* **2009**, *374*, 491–499. [CrossRef]

4. Rotge, J.Y.; Guehl, D.; Dilharreguy, B.; Cuny, E.; Tignol, J.; Bioulac, B.; Allard, M.; Burbaud, P.; Aouizerate, B. Provocation of obsessive-compulsive symptoms: A quantitative voxel-based meta-analysis of functional neuroimaging studies. *J. Psychiatry Neurosci.* **2008**, *33*, 405–412. [PubMed]

5. Haynes, W.I.; Mallet, L. High-frequency stimulation of deep brain structures in obsessive-compulsive disorder: The search for a valid circuit. *Eur. J. Neurosci.* **2010**, *32*, 1118–1127. [CrossRef] [PubMed]

6. Mallet, L.; Polosan, M.; Jaafari, N.; Baup, N.; Welter, M.L.; Fontaine, D.; du Montcel, S.T.; Yelnik, J.; Chereau, I.; Arbus, C.; et al. Subthalamic nucleus stimulation in severe obsessive-compulsive disorder. *N. Engl. J. Med.* **2008**, *359*, 2121–2134. [CrossRef] [PubMed]

7. Burguiere, E.; Monteiro, P.; Feng, G.; Graybiel, A.M. Optogenetic stimulation of lateral orbitofronto-striatal pathway suppresses compulsive behaviors. *Science* **2013**, *340*, 1243–1246. [CrossRef] [PubMed]

8. Costas, J.; Carrera, N.; Alonso, P.; Gurriaran, X.; Segalas, C.; Real, E.; Lopez-Sola, C.; Mas, S.; Gasso, P.; Domenech, L.; et al. Exon-focused genome-wide association study of obsessive-compulsive disorder and shared polygenic risk with schizophrenia. *Transl. Psychiatry* **2016**, *6*, e768. [CrossRef] [PubMed]

9. Cappi, C.; Brentani, H.; Lima, L.; Sanders, S.J.; Zai, G.; Diniz, B.J.; Reis, V.N.; Hounie, A.G.; Conceicao do Rosario, M.; Mariani, D.; et al. Whole-exome sequencing in obsessive-compulsive disorder identifies rare mutations in immunological and neurodevelopmental pathways. *Transl. Psychiatry* **2016**, *6*, e764. [CrossRef] [PubMed]

10. Rodriguez, N.; Morer, A.; Gonzalez-Navarro, E.A.; Gasso, P.; Boloc, D.; Serra-Pages, C.; Lafuente, A.; Lazaro, L.; Mas, S. Human-leukocyte antigen class ii genes in early-onset obsessive-compulsive disorder. *World J. Biol. Psychiatry* **2017**. [CrossRef] [PubMed]

11. Frick, L.; Pittenger, C. Microglial dysregulation in OCD, tourette syndrome, and PANDAS. *J. Immunol. Res.* **2016**, *2016*, 8606057. [CrossRef] [PubMed]

12. Dale, R.C.; Merheb, V.; Pillai, S.; Wang, D.; Cantrill, L.; Murphy, T.K.; Ben-Pazi, H.; Varadkar, S.; Aumann, T.D.; Horne, M.K.; et al. Antibodies to surface dopamine-2 receptor in autoimmune movement and psychiatric disorders. *Brain* **2012**, *135*, 3453–3468. [CrossRef] [PubMed]

13. Swedo, S.E.; Leckman, J.F.; Rose, N.R. From research subgroup to clinical syndrome: Modifying the PANDAS criteria to describe PANS (pediatric acute-onset neuropsychiatric syndrome). *Pediatr. Ther.* **2012**, *2*, 1000113. [CrossRef]

14. Sutterland, A.L.; Fond, G.; Kuin, A.; Koeter, M.W.; Lutter, R.; van Gool, T.; Yolken, R.; Szoke, A.; Leboyer, M.; de Haan, L. Beyond the association. *Toxoplasma gondii* in schizophrenia, bipolar disorder, and addiction: Systematic review and meta-analysis. *Acta Psychiatr. Scand.* **2015**, *132*, 161–179. [CrossRef] [PubMed]

15. Baldermann, J.C.; Schuller, T.; Huys, D.; Becker, I.; Timmermann, L.; Jessen, F.; Visser-Vandewalle, V.; Kuhn, J. Deep brain stimulation for tourette-syndrome: A systematic review and meta-analysis. *Brain Stimul.* **2016**, *9*, 296–304. [CrossRef] [PubMed]

16. Kisely, S.; Hall, K.; Siskind, D.; Frater, J.; Olson, S.; Crompton, D. Deep brain stimulation for obsessive-compulsive disorder: A systematic review and meta-analysis. *Psychol. Med.* **2014**, *44*, 3533–3542. [CrossRef] [PubMed]

17. Blomstedt, P.; Sjoberg, R.L.; Hansson, M.; Bodlund, O.; Hariz, M.I. Deep brain stimulation in the treatment of obsessive-compulsive disorder. *World Neurosurg.* **2013**, *80*, e245–e253. [CrossRef] [PubMed]

18. Hirschtritt, M.E.; Bloch, M.H.; Mathews, C.A. Obsessive-compulsive disorder: Advances in diagnosis and treatment. *JAMA* **2017**, *317*, 1358–1367. [CrossRef] [PubMed]

19. Rose-John, S. Interleukin-6 family cytokines. *Cold Spring Harb. Perspect. Biol.* **2018**, *10*. [CrossRef] [PubMed]

20. Weizman, R.; Laor, N.; Barber, Y.; Hermesh, H.; Notti, I.; Djaldetti, M.; Bessler, H. Cytokine production in obsessive-compulsive disorder. *Biol. Psychiatry* **1996**, *40*, 908–912. [CrossRef]

21. Maes, M.; Meltzer, H.Y.; Bosmans, E. Psychoimmune investigation in obsessive-compulsive disorder: Assays of plasma transferrin, IL-2 and IL-6 receptor, and IL-1 β and IL-6 concentrations. *Neuropsychobiology* **1994**, *30*, 57–60. [CrossRef] [PubMed]

22. Cappi, C.; Muniz, R.K.; Sampaio, A.S.; Cordeiro, Q.; Brentani, H.; Palacios, S.A.; Marques, A.H.; Vallada, H.; Miguel, E.C.; Guilherme, L.; et al. Association study between functional polymorphisms in the tnf-alpha gene and obsessive-compulsive disorder. *Arq. Neuropsiquiatr.* **2012**, *70*, 87–90. [CrossRef] [PubMed]

23. Fontenelle, L.F.; Barbosa, I.G.; Luna, J.V.; de Sousa, L.P.; Abreu, M.N.; Teixeira, A.L. A cytokine study of adult patients with obsessive-compulsive disorder. *Compr. Psychiatry* **2012**, *53*, 797–804. [CrossRef] [PubMed]

24. Fluitman, S.B.; Denys, D.A.; Heijnen, C.J.; Westenberg, H.G. Disgust affects TNF-alpha, IL-6 and noradrenalin levels in patients with obsessive-compulsive disorder. *Psychoneuroendocrinology* **2010**, *35*, 906–911. [CrossRef] [PubMed]

25. Fluitman, S.; Denys, D.; Vulink, N.; Schutters, S.; Heijnen, C.; Westenberg, H. Lipopolysaccharide-induced cytokine production in obsessive-compulsive disorder and generalized social anxiety disorder. *Psychiatry Res.* **2010**, *178*, 313–316. [CrossRef] [PubMed]

26. Hounie, A.G.; Cappi, C.; Cordeiro, Q.; Sampaio, A.S.; Moraes, I.; Rosario, M.C.; Palacios, S.A.; Goldberg, A.C.; Vallada, H.P.; Machado-Lima, A.; et al. Tnf-alpha polymorphisms are associated with obsessive-compulsive disorder. *Neurosci. Lett.* **2008**, *442*, 86–90. [CrossRef] [PubMed]

27. Konuk, N.; Tekin, I.O.; Ozturk, U.; Atik, L.; Atasoy, N.; Bektas, S.; Erdogan, A. Plasma levels of tumor necrosis factor-α and interleukin-6 in obsessive compulsive disorder. *Mediat. Inflamm.* **2007**, *2007*, 65704. [CrossRef] [PubMed]

28. Denys, D.; Fluitman, S.; Kavelaars, A.; Heijnen, C.; Westenberg, H. Decreased TNF-alpha and NK activity in obsessive-compulsive disorder. *Psychoneuroendocrinology* **2004**, *29*, 945–952. [CrossRef] [PubMed]

29. Monteleone, P.; Catapano, F.; Fabrazzo, M.; Tortorella, A.; Maj, M. Decreased blood levels of tumor necrosis factor-alpha in patients with obsessive-compulsive disorder. *Neuropsychobiology* **1998**, *37*, 182–185. [CrossRef] [PubMed]

30. Brambilla, F.; Perna, G.; Bellodi, L.; Arancio, C.; Bertani, A.; Perini, G.; Carraro, C.; Gava, F. Plasma interleukin-1 beta and tumor necrosis factor concentrations in obsessive-compulsive disorders. *Biol. Psychiatry* **1997**, *42*, 976–981. [CrossRef]

31. Gray, S.M.; Bloch, M.H. Systematic review of proinflammatory cytokines in obsessive-compulsive disorder. *Curr. Psychiatry Rep.* **2012**, *14*, 220–228. [CrossRef] [PubMed]

32. Colak Sivri, R.; Bilgic, A.; Kilinc, I. Cytokine, chemokine and bdnf levels in medication-free pediatric patients with obsessive-compulsive disorder. *Eur. Child Adolesc. Psychiatry* **2018**, *27*, 977–984. [CrossRef] [PubMed]

33. Simsek, S.; Yuksel, T.; Cim, A.; Kaya, S. Serum cytokine profiles of children with obsessive-compulsive disorder shows the evidence of autoimmunity. *Int. J. Neuropsychopharmacol.* **2016**, *19*, pyw027. [CrossRef] [PubMed]

34. Rao, N.P.; Venkatasubramanian, G.; Ravi, V.; Kalmady, S.; Cherian, A.; Yc, J.R. Plasma cytokine abnormalities in drug-naive, comorbidity-free obsessive-compulsive disorder. *Psychiatry Res.* **2015**, *229*, 949–952. [CrossRef] [PubMed]

35. Rodriguez, N.; Morer, A.; Gonzalez-Navarro, E.A.; Serra-Pages, C.; Boloc, D.; Torres, T.; Garcia-Cerro, S.; Mas, S.; Gasso, P.; Lazaro, L. Inflammatory dysregulation of monocytes in pediatric patients with obsessive-compulsive disorder. *J. Neuroinflamm.* **2017**, *14*, 261. [CrossRef] [PubMed]

36. Jiang, C.; Ma, X.; Qi, S.; Han, G.; Li, Y.; Liu, Y.; Liu, L. Association between tnf-alpha-238g/a gene polymorphism and ocd susceptibility: A meta-analysis. *Medicine* **2018**, *97*, e9769. [CrossRef] [PubMed]

37. Uguz, F.; Onder Sonmez, E.; Sahingoz, M.; Gokmen, Z.; Basaran, M.; Gezginc, K.; Sonmez, G.; Kaya, N.; Yilmaz, E.; Erdem, S.S.; et al. Neuroinflammation in the fetus exposed to maternal obsessive-compulsive disorder during pregnancy: A comparative study on cord blood tumor necrosis factor-alpha levels. *Compr. Psychiatry* **2014**, *55*, 861–865. [CrossRef] [PubMed]

38. Bo, Y.; Liu, S.; Yin, Y.; Wang, Z.; Cui, J.; Zong, J.; Zhang, X.; Li, X. Association study between IL-1β-511 C/T polymorphism and obsessive-compulsive disorder (OCD) in chinese han population. *Int. J. Psychiatry Med.* **2013**, *46*, 145–152. [CrossRef] [PubMed]

39. Zhang, X.; Yin, Y.; Liu, S.; Ma, X. A case-control association study between obsessive-compulsive disorder (OCD) and the MCP-1-2518G/A polymorphism in a chinese sample. *Rev. Bras. Psiquiatr.* **2012**, *34*, 451–453. [CrossRef] [PubMed]

40. Liu, S.; Liu, Y.; Zhang, X.; Ma, X. Lack of association of -251T/A polymorphism in interleukin 8 gene with susceptibility to obsessive-compulsive disorder in Chinese Han population. *Cytokine* **2012**, *59*, 209–210. [CrossRef] [PubMed]

41. Carpenter, L.L.; Heninger, G.R.; McDougle, C.J.; Tyrka, A.R.; Epperson, C.N.; Price, L.H. Cerebrospinal fluid interleukin-6 in obsessive-compulsive disorder and trichotillomania. *Psychiatry Res.* **2002**, *112*, 257–262. [CrossRef]

42. Akdis, M.; Aab, A.; Altunbulakli, C.; Azkur, K.; Costa, R.A.; Crameri, R.; Duan, S.; Eiwegger, T.; Eljaszewicz, A.; Ferstl, R.; et al. Interleukins (from IL-1 to IL-38), interferons, transforming growth factor beta, and TNF-alpha: Receptors, functions, and roles in diseases. *J. Allergy Clin. Immunol.* **2016**, *138*, 984–1010. [CrossRef] [PubMed]

43. Yuce, M.; Guner, S.N.; Karabekiroglu, K.; Baykal, S.; Kilic, M.; Sancak, R.; Karabekiroglu, A. Association of tourette syndrome and obsessive-compulsive disorder with allergic diseases in children and adolescents: A preliminary study. *Eur. Rev. Med. Pharmacol. Sci.* **2014**, *18*, 303–310. [PubMed]

44. Stellwagen, D.; Malenka, R.C. Synaptic scaling mediated by glial tnf-alpha. *Nature* **2006**, *440*, 1054–1059. [CrossRef] [PubMed]

45. Beattie, E.C.; Stellwagen, D.; Morishita, W.; Bresnahan, J.C.; Ha, B.K.; Von Zastrow, M.; Beattie, M.S.; Malenka, R.C. Control of synaptic strength by glial tnfalpha. *Science* **2002**, *295*, 2282–2285. [CrossRef] [PubMed]

46. Krabbe, G.; Minami, S.S.; Etchegaray, J.I.; Taneja, P.; Djukic, B.; Davalos, D.; Le, D.; Lo, I.; Zhan, L.; Reichert, M.C.; et al. Microglial nfkappab-tnfalpha hyperactivation induces obsessive-compulsive behavior in mouse models of progranulin-deficient frontotemporal dementia. *Proc. Natl. Acad. Sci. USA* **2017**, *114*, 5029–5034. [CrossRef] [PubMed]

47. Gruol, D.L. Il-6 regulation of synaptic function in the cns. *Neuropharmacology* **2015**, *96*, 42–54. [CrossRef] [PubMed]

48. Bhattacharyya, S.; Khanna, S.; Chakrabarty, K.; Mahadevan, A.; Christopher, R.; Shankar, S.K. Anti-brain autoantibodies and altered excitatory neurotransmitters in obsessive-compulsive disorder. *Neuropsychopharmacology* **2009**, *34*, 2489–2496. [CrossRef] [PubMed]

49. Morer, A.; Lazaro, L.; Sabater, L.; Massana, J.; Castro, J.; Graus, F. Antineuronal antibodies in a group of children with obsessive-compulsive disorder and tourette syndrome. *J. Psychiatr. Res.* **2008**, *42*, 64–68. [CrossRef] [PubMed]

50. Dale, R.C.; Heyman, I.; Giovannoni, G.; Church, A.W. Incidence of anti-brain antibodies in children with obsessive-compulsive disorder. *Br. J. Psychiatry* **2005**, *187*, 314–319. [CrossRef] [PubMed]

51. Pearlman, D.M.; Vora, H.S.; Marquis, B.G.; Najjar, S.; Dudley, L.A. Anti-basal ganglia antibodies in primary obsessive-compulsive disorder: Systematic review and meta-analysis. *Br. J. Psychiatry* **2014**, *205*, 8–16. [CrossRef] [PubMed]

52. Akaltun, I.; Kara, S.S.; Kara, T. The relationship between *Toxoplasma gondii* IgG antibodies and generalized anxiety disorder and obsessive-compulsive disorder in children and adolescents: A new approach. *Nord J. Psychiatry* **2018**, *72*, 57–62. [CrossRef] [PubMed]

53. Mataix-Cols, D.; Frans, E.; Perez-Vigil, A.; Kuja-Halkola, R.; Gromark, C.; Isomura, K.; Fernandez de la Cruz, L.; Serlachius, E.; Leckman, J.F.; Crowley, J.J.; et al. A total-population multigenerational family clustering study of autoimmune diseases in obsessive-compulsive disorder and tourette's/chronic tic disorders. *Mol. Psychiatry* **2017**. [CrossRef] [PubMed]

54. Flegr, J.; Horacek, J. Toxoplasma-infected subjects report an obsessive-compulsive disorder diagnosis more often and score higher in obsessive-compulsive inventory. *Eur. Psychiatry* **2017**, *40*, 82–87. [CrossRef] [PubMed]

55. Nicolini, H.; Lopez, Y.; Genis-Mendoza, A.D.; Manrique, V.; Lopez-Canovas, L.; Niubo, E.; Hernandez, L.; Bobes, M.A.; Riveron, A.M.; Lopez-Casamichana, M.; et al. Detection of anti-streptococcal, antienolase, and anti-neural antibodies in subjects with early-onset psychiatric disorders. *Actas Esp. Psiquiatr.* **2015**, *43*, 35–41. [PubMed]

56. Singer, H.S.; Mascaro-Blanco, A.; Alvarez, K.; Morris-Berry, C.; Kawikova, I.; Ben-Pazi, H.; Thompson, C.B.; Ali, S.F.; Kaplan, E.L.; Cunningham, M.W. Neuronal antibody biomarkers for sydenham's chorea identify a new group of children with chronic recurrent episodic acute exacerbations of tic and obsessive compulsive symptoms following a streptococcal infection. *PLoS ONE* **2015**, *10*, e0120499. [CrossRef] [PubMed]

57. Frankovich, J.; Thienemann, M.; Pearlstein, J.; Crable, A.; Brown, K.; Chang, K. Multidisciplinary clinic dedicated to treating youth with pediatric acute-onset neuropsychiatric syndrome: Presenting characteristics of the first 47 consecutive patients. *J. Child Adolesc. Psychopharmacol.* **2015**, *25*, 38–47. [CrossRef] [PubMed]

58. Cox, C.J.; Zuccolo, A.J.; Edwards, E.V.; Mascaro-Blanco, A.; Alvarez, K.; Stoner, J.; Chang, K.; Cunningham, M.W. Antineuronal antibodies in a heterogeneous group of youth and young adults with tics and obsessive-compulsive disorder. *J. Child Adolesc. Psychopharmacol.* **2015**, *25*, 76–85. [CrossRef] [PubMed]

59. Ebrahimi Taj, F.; Noorbakhsh, S.; Ghavidel Darestani, S.; Shirazi, E.; Javadinia, S. Group a beta-hemolytic streptococcal infection in children and the resultant neuro-psychiatric disorder; A cross sectional study; Tehran, iran. *Basic Clin. Neurosci.* **2015**, *6*, 38–43. [PubMed]

60. Murphy, T.K.; Patel, P.D.; McGuire, J.F.; Kennel, A.; Mutch, P.J.; Parker-Athill, E.C.; Hanks, C.E.; Lewin, A.B.; Storch, E.A.; Toufexis, M.D.; et al. Characterization of the pediatric acute-onset neuropsychiatric syndrome phenotype. *J. Child Adolesc. Psychopharmacol.* **2015**, *25*, 14–25. [CrossRef] [PubMed]

61. Murphy, T.K.; Storch, E.A.; Lewin, A.B.; Edge, P.J.; Goodman, W.K. Clinical factors associated with pediatric autoimmune neuropsychiatric disorders associated with streptococcal infections. *J. Pediatr.* **2012**, *160*, 314–319. [CrossRef] [PubMed]

62. Leckman, J.F.; King, R.A.; Gilbert, D.L.; Coffey, B.J.; Singer, H.S.; Dure, L.S.T.; Grantz, H.; Katsovich, L.; Lin, H.; Lombroso, P.J.; et al. Streptococcal upper respiratory tract infections and exacerbations of tic and obsessive-compulsive symptoms: A prospective longitudinal study. *J. Am. Acad. Child Adolesc. Psychiatry* **2011**, *50*, 108–118.e3. [CrossRef] [PubMed]

63. Miman, O.; Mutlu, E.A.; Ozcan, O.; Atambay, M.; Karlidag, R.; Unal, S. Is there any role of *Toxoplasma gondii* in the etiology of obsessive-compulsive disorder? *Psychiatry Res.* **2010**, *177*, 263–265. [CrossRef] [PubMed]

64. Gause, C.; Morris, C.; Vernekar, S.; Pardo-Villamizar, C.; Grados, M.A.; Singer, H.S. Antineuronal antibodies in OCD: Comparisons in children with OCD-only, OCD+Chronic tics and OCD+PANDAS. *J. Neuroimmunol.* **2009**, *214*, 118–124. [CrossRef] [PubMed]

65. Kirvan, C.A.; Swedo, S.E.; Snider, L.A.; Cunningham, M.W. Antibody-mediated neuronal cell signaling in behavior and movement disorders. *J. Neuroimmunol.* **2006**, *179*, 173–179. [CrossRef] [PubMed]

66. Morer, A.; Vinas, O.; Lazaro, L.; Calvo, R.; Andres, S.; Bosch, J.; Gasto, C.; Massana, J.; Castro, J. Subtyping obsessive-compulsive disorder: Clinical and immunological findings in child and adult onset. *J. Psychiatr. Res.* **2006**, *40*, 207–213. [CrossRef] [PubMed]

67. Singer, H.S.; Hong, J.J.; Yoon, D.Y.; Williams, P.N. Serum autoantibodies do not differentiate PANDAS and tourette syndrome from controls. *Neurology* **2005**, *65*, 1701–1707. [CrossRef] [PubMed]

68. Pavone, P.; Bianchini, R.; Parano, E.; Incorpora, G.; Rizzo, R.; Mazzone, L.; Trifiletti, R.R. Anti-brain antibodies in PANDAS versus uncomplicated streptococcal infection. *Pediatr. Neurol.* **2004**, *30*, 107–110. [CrossRef]

69. Murphy, T.K.; Sajid, M.; Soto, O.; Shapira, N.; Edge, P.; Yang, M.; Lewis, M.H.; Goodman, W.K. Detecting pediatric autoimmune neuropsychiatric disorders associated with streptococcus in children with obsessive-compulsive disorder and tics. *Biol. Psychiatry* **2004**, *55*, 61–68. [CrossRef]

70. Luo, F.; Leckman, J.F.; Katsovich, L.; Findley, D.; Grantz, H.; Tucker, D.M.; Lombroso, P.J.; King, R.A.; Bessen, D.E. Prospective longitudinal study of children with tic disorders and/or obsessive-compulsive disorder: Relationship of symptom exacerbations to newly acquired streptococcal infections. *Pediatrics* **2004**, *113*, e578–e585. [CrossRef] [PubMed]

71. Inoff-Germain, G.; Rodriguez, R.S.; Torres-Alcantara, S.; Diaz-Jimenez, M.J.; Swedo, S.E.; Rapoport, J.L. An immunological marker (D8/17) associated with rheumatic fever as a predictor of childhood psychiatric disorders in a community sample. *J. Child Psychol. Psychiatry* **2003**, *44*, 782–790. [CrossRef] [PubMed]

72. Murphy, M.L.; Pichichero, M.E. Prospective identification and treatment of children with pediatric autoimmune neuropsychiatric disorder associated with group a streptococcal infection (PANDAS). *Arch. Pediatr. Adolesc. Med.* **2002**, *156*, 356–361. [CrossRef] [PubMed]

73. Eisen, J.L.; Leonard, H.L.; Swedo, S.E.; Price, L.H.; Zabriskie, J.B.; Chiang, S.Y.; Karitani, M.; Rasmussen, S.A. The use of antibody D8/17 to identify b cells in adults with obsessive-compulsive disorder. *Psychiatry Res.* **2001**, *104*, 221–225. [CrossRef]

74. Murphy, T.K.; Benson, N.; Zaytoun, A.; Yang, M.; Braylan, R.; Ayoub, E.; Goodman, W.K. Progress toward analysis of D8/17 binding to b cells in children with obsessive compulsive disorder and/or chronic tic disorder. *J. Neuroimmunol.* **2001**, *120*, 146–151. [CrossRef]

75. Peterson, B.S.; Leckman, J.F.; Tucker, D.; Scahill, L.; Staib, L.; Zhang, H.; King, R.; Cohen, D.J.; Gore, J.C.; Lombroso, P. Preliminary findings of antistreptococcal antibody titers and basal ganglia volumes in tic, obsessive-compulsive, and attention deficit/hyperactivity disorders. *Arch. Gen. Psychiatry* **2000**, *57*, 364–372. [CrossRef] [PubMed]

76. Marazziti, D.; Presta, S.; Pfanner, C.; Gemignani, A.; Rossi, A.; Sbrana, S.; Rocchi, V.; Ambrogi, F.; Cassano, G.B. Immunological alterations in adult obsessive-compulsive disorder. *Biol. Psychiatry* **1999**, *46*, 810–814. [CrossRef]

77. Chapman, F.; Visvanathan, K.; Carreno-Manjarrez, R.; Zabriskie, J.B. A flow cytometric assay for d8/17 b cell marker in patients with tourette's syndrome and obsessive compulsive disorder. *J. Immunol. Methods* **1998**, *219*, 181–186. [CrossRef]

78. Khanna, S.; Ravi, V.; Shenoy, P.K.; Chandramukhi, A.; Channabasavanna, S.M. Viral antibodies in blood in obsessive compulsive disorder. *Indian J. Psychiatry* **1997**, *39*, 190–195. [PubMed]

79. Khanna, S.; Ravi, V.; Shenoy, P.K.; Chandramuki, A.; Channabasavanna, S.M. Cerebrospinal fluid viral antibodies in obsessive-compulsive disorder in an indian population. *Biol. Psychiatry* **1997**, *41*, 883–890. [CrossRef]

80. Murphy, T.K.; Goodman, W.K.; Fudge, M.W.; Williams, R.C., Jr.; Ayoub, E.M.; Dalal, M.; Lewis, M.H.; Zabriskie, J.B. B lymphocyte antigen D8/17: A peripheral marker for childhood-onset obsessive-compulsive disorder and tourette's syndrome? *Am. J. Psychiatry* **1997**, *154*, 402–407. [PubMed]

81. Swedo, S.E.; Leonard, H.L.; Mittleman, B.B.; Allen, A.J.; Rapoport, J.L.; Dow, S.P.; Kanter, M.E.; Chapman, F.; Zabriskie, J. Identification of children with pediatric autoimmune neuropsychiatric disorders associated with streptococcal infections by a marker associated with rheumatic fever. *Am. J. Psychiatry* **1997**, *154*, 110–112. [PubMed]

82. Zabriskie, J.B.; Lavenchy, D.; Williams, R.C., Jr.; Fu, S.M.; Yeadon, C.A.; Fotino, M.; Braun, D.G. Rheumatic fever-associated B cell alloantigens as identified by monoclonal antibodies. *Arthritis Rheum.* **1985**, *28*, 1047–1051. [CrossRef] [PubMed]

83. Patarroyo, M.E.; Winchester, R.J.; Vejerano, A.; Gibofsky, A.; Chalem, F.; Zabriskie, J.B.; Kunkel, H.G. Association of a B-cell alloantigen with susceptibility to rheumatic fever. *Nature* **1979**, *278*, 173–174. [CrossRef] [PubMed]

84. Morer, A.; Vinas, O.; Lazaro, L.; Bosch, J.; Toro, J.; Castro, J. D8/17 monoclonal antibody: An unclear neuropsychiatric marker. *Behav. Neurol.* **2005**, *16*, 1–8. [CrossRef] [PubMed]

85. Atmaca, M.; Kilic, F.; Koseoglu, F.; Ustundag, B. Neutrophils are decreased in obsessive-compulsive disorder: Preliminary investigation. *Psychiatry Investig.* **2011**, *8*, 362–365. [CrossRef] [PubMed]

86. Barber, Y.; Toren, P.; Achiron, A.; Noy, S.; Wolmer, L.; Weizman, R.; Laor, N. T cell subsets in obsessive-compulsive disorder. *Neuropsychobiology* **1996**, *34*, 63–66. [CrossRef] [PubMed]

87. Denys, D.; Fluitman, S.; Kavelaars, A.; Heijnen, C.; Westenberg, H.G. Effects of paroxetine and venlafaxine on immune parameters in patients with obsessive compulsive disorder. *Psychoneuroendocrinology* **2006**, *31*, 355–360. [CrossRef] [PubMed]

88. Marazziti, D.; Mungai, F.; Masala, I.; Baroni, S.; Vivarelli, L.; Ambrogi, F.; Catena Dell'Osso, M.; Consoli, G.; Massimetti, G.; Dell'Osso, L. Normalisation of immune cell imbalance after pharmacological treatments of patients suffering from obsessive-compulsive disorder. *J. Psychopharmacol.* **2009**, *23*, 567–573. [CrossRef] [PubMed]

89. Marazziti, D.; Baroni, S.; Masala, I.; Giannaccini, G.; Mungai, F.; Di Nasso, E.; Cassano, G.B. Decreased lymphocyte 3h-paroxetine binding in obsessive-compulsive disorder. *Neuropsychobiology* **2003**, *47*, 128–130. [CrossRef] [PubMed]

90. Marazziti, D.; Ori, M.; Nardini, M.; Rossi, A.; Nardi, I.; Cassano, G.B. Mrna expression of serotonin receptors of type 2c and 5a in human resting lymphocytes. *Neuropsychobiology* **2001**, *43*, 123–126. [CrossRef] [PubMed]

91. Rocca, P.; Beoni, A.M.; Eva, C.; Ferrero, P.; Maina, G.; Bogetto, F.; Ravizza, L. Lymphocyte peripheral benzodiazepine receptor mrna decreases in obsessive-compulsive disorder. *Eur. Neuropsychopharmacol.* **2000**, *10*, 337–340. [CrossRef]

92. Ravindran, A.V.; Griffiths, J.; Merali, Z.; Anisman, H. Circulating lymphocyte subsets in obsessive compulsive disorder, major depression and normal controls. *J. Affect. Disords* **1999**, *52*, 1–10. [CrossRef]

93. Rocca, P.; Ferrero, P.; Gualerzi, A.; Zanalda, E.; Maina, G.; Bergamasco, B.; Ravizza, L. Peripheral-type benzodiazepine receptors in anxiety disorders. *Acta Psychiatr. Scand.* **1991**, *84*, 537–544. [CrossRef] [PubMed]

94. Ursoiu, F.; Moleriu, L.; Lungeanu, D.; Puschita, M. The association between hiv clinical disease severity and psychiatric disorders as seen in western romania. *AIDS Care* **2018**. [CrossRef] [PubMed]

95. Dale, R.C.; Heyman, I.; Surtees, R.A.; Church, A.J.; Giovannoni, G.; Goodman, R.; Neville, B.G. Dyskinesias and associated psychiatric disorders following streptococcal infections. *Arch. Dis. Child.* **2004**, *89*, 604–610. [CrossRef] [PubMed]

96. Giulino, L.; Gammon, P.; Sullivan, K.; Franklin, M.; Foa, E.; Maid, R.; March, J.S. Is parental report of upper respiratory infection at the onset of obsessive-compulsive disorder suggestive of pediatric autoimmune neuropsychiatric disorder associated with streptococcal infection? *J. Child Adolesc. Psychopharmacol.* **2002**, *12*, 157–164. [CrossRef] [PubMed]

97. Lougee, L.; Perlmutter, S.J.; Nicolson, R.; Garvey, M.A.; Swedo, S.E. Psychiatric disorders in first-degree relatives of children with pediatric autoimmune neuropsychiatric disorders associated with streptococcal infections (PANDAS). *J. Am. Acad. Child Adolesc. Psychiatry* **2000**, *39*, 1120–1126. [CrossRef] [PubMed]

98. Johnco, C.; Kugler, B.B.; Murphy, T.K.; Storch, E.A. Obsessive-compulsive symptoms in adults with lyme disease. *Gen. Hosp. Psychiatry* **2018**, *51*, 85–89. [CrossRef] [PubMed]

99. Stagi, S.; Lepri, G.; Rigante, D.; Matucci Cerinic, M.; Falcini, F. Cross-sectional evaluation of plasma vitamin d levels in a large cohort of italian patients with pediatric autoimmune neuropsychiatric disorders associated with streptococcal infections. *J. Child Adolesc. Psychopharmacol.* **2018**, *28*, 124–129. [CrossRef] [PubMed]

100. Calaprice, D.; Tona, J.; Parker-Athill, E.C.; Murphy, T.K. A survey of pediatric acute-onset neuropsychiatric syndrome characteristics and course. *J. Child Adolesc. Psychopharmacol.* **2017**, *27*, 607–618. [CrossRef] [PubMed]

101. Wang, H.C.; Lau, C.I.; Lin, C.C.; Chang, A.; Kao, C.H. Group a streptococcal infections are associated with increased risk of pediatric neuropsychiatric disorders: A taiwanese population-based cohort study. *J. Clin. Psychiatry* **2016**, *77*, e848–e854. [CrossRef] [PubMed]

102. Murphy, T.K.; Storch, E.A.; Turner, A.; Reid, J.M.; Tan, J.; Lewin, A.B. Maternal history of autoimmune disease in children presenting with tics and/or obsessive-compulsive disorder. *J. Neuroimmunol.* **2010**, *229*, 243–247. [CrossRef] [PubMed]

103. Kurlan, R.; Johnson, D.; Kaplan, E.L.; Tourette Syndrome Study Group. Streptococcal infection and exacerbations of childhood tics and obsessive-compulsive symptoms: A prospective blinded cohort study. *Pediatrics* **2008**, *121*, 1188–1197. [CrossRef] [PubMed]

104. Giedd, J.N.; Rapoport, J.L.; Garvey, M.A.; Perlmutter, S.; Swedo, S.E. Mri assessment of children with obsessive-compulsive disorder or tics associated with streptococcal infection. *Am. J. Psychiatry* **2000**, *157*, 281–283. [CrossRef] [PubMed]

105. Swedo, S.E.; Leonard, H.L.; Garvey, M.; Mittleman, B.; Allen, A.J.; Perlmutter, S.; Lougee, L.; Dow, S.; Zamkoff, J.; Dubbert, B.K. Pediatric autoimmune neuropsychiatric disorders associated with streptococcal infections: Clinical description of the first 50 cases. *Am. J. Psychiatry* **1998**, *155*, 264–271. [CrossRef] [PubMed]

106. *Streptococcus Pyogenes: Basic Biology to Clinical Manifestations*; Ferretti, J.J.; Stevens, D.L.; Fischetti, V.A. (Eds.) University of Oklahoma Health Sciences Center: Oklahoma City, OK, USA, 2016.

107. Cunningham, M.W. Post-streptococcal autoimmune sequelae: Rheumatic fever and beyond. In *Streptococcus Pyogenes: Basic Biology to Clinical Manifestations*; Ferretti, J.J., Stevens, D.L., Fischetti, V.A., Eds.; University of Oklahoma Health Sciences Center: Oklahoma City, OK, USA, 2016.

108. Dajani, A.S. Guidelines for the diagnosis of rheumatic fever. Jones criteria, 1992 update. Special writing group of the committee on rheumatic fever, endocarditis, and kawasaki disease of the council on cardiovascular disease in the young of the american heart association. *JAMA* **1992**, *268*, 2069–2073.

109. Yanagawa, B.; Butany, J.; Verma, S. Update on rheumatic heart disease. *Curr. Opin. Cardiol.* **2016**, *31*, 162–168. [CrossRef] [PubMed]

110. Macerollo, A.; Martino, D. Pediatric autoimmune neuropsychiatric disorders associated with streptococcal infections (PANDAS): An evolving concept. *Tremor Other Hyperkinet. Mov.* **2013**, *3*. [CrossRef]

111. Cunningham, M.W.; Cox, C.J. Autoimmunity against dopamine receptors in neuropsychiatric and movement disorders: A review of sydenham chorea and beyond. *Acta Physiol.* **2016**, *216*, 90–100. [CrossRef] [PubMed]

112. Park, J. Movement disorders following cerebrovascular lesion in the basal ganglia circuit. *J. Mov. Disords* **2016**, *9*, 71–79. [CrossRef] [PubMed]

113. Orefici, G.; Cardona, F.; Cox, C.J.; Cunningham, M.W. Pediatric autoimmune neuropsychiatric disorders associated with streptococcal infections (PANDAS). In *Streptococcus Pyogenes: Basic Biology to Clinical Manifestations*; Ferretti, J.J., Stevens, D.L., Fischetti, V.A., Eds.; University of Oklahoma Health Sciences Center: Oklahoma City, OK, USA, 2016.

114. Cox, C.J.; Sharma, M.; Leckman, J.F.; Zuccolo, J.; Zuccolo, A.; Kovoor, A.; Swedo, S.E.; Cunningham, M.W. Brain human monoclonal autoantibody from sydenham chorea targets dopaminergic neurons in transgenic mice and signals dopamine d2 receptor: Implications in human disease. *J. Immunol.* **2013**, *191*, 5524–5541. [CrossRef] [PubMed]

115. Fibbe, L.A.; Cath, D.C.; van den Heuvel, O.A.; Veltman, D.J.; Tijssen, M.A.; van Balkom, A.J. Relationship between movement disorders and obsessive-compulsive disorder: Beyond the obsessive-compulsive-tic phenotype. A systematic review. *J. Neurol. Neurosurg. Psychiatry* **2012**, *83*, 646–654. [CrossRef] [PubMed]

116. Frick, L.R.; Rapanelli, M.; Jindachomthong, K.; Grant, P.; Leckman, J.F.; Swedo, S.; Williams, K.; Pittenger, C. Differential binding of antibodies in PANDAS patients to cholinergic interneurons in the striatum. *Brain. Behav. Immun.* **2018**, *69*, 304–311. [CrossRef] [PubMed]

117. Jaspers-Fayer, F.; Han, S.H.J.; Chan, E.; McKenney, K.; Simpson, A.; Boyle, A.; Ellwyn, R.; Stewart, S.E. Prevalence of acute-onset subtypes in pediatric obsessive-compulsive disorder. *J. Child Adolesc. Psychopharmacol.* **2017**, *27*, 332–341. [CrossRef] [PubMed]

118. Leon, J.; Hommer, R.; Grant, P.; Farmer, C.; D'Souza, P.; Kessler, R.; Williams, K.; Leckman, J.F.; Swedo, S. Longitudinal outcomes of children with pediatric autoimmune neuropsychiatric disorder associated with streptococcal infections (PANDAS). *Eur. Child Adolesc. Psychiatry* **2018**, *27*, 637–643. [CrossRef] [PubMed]

119. Skoog, G.; Skoog, I. A 40-year follow-up of patients with obsessive-compulsive disorder. *Arch. Gen. Psychiatry* **1999**, *56*, 121–127. [CrossRef] [PubMed]

120. Calaprice, D.; Tona, J.; Murphy, T.K. Treatment of pediatric acute-onset neuropsychiatric disorder in a large survey population. *J. Child Adolesc. Psychopharmacol.* **2018**, *28*, 92–103. [CrossRef] [PubMed]

121. Brown, K.; Farmer, C.; Farhadian, B.; Hernandez, J.; Thienemann, M.; Frankovich, J. Pediatric acute-onset neuropsychiatric syndrome response to oral corticosteroid bursts: An observational study of patients in an academic community-based PANS clinic. *J. Child Adolesc. Psychopharmacol.* **2017**, *27*, 629–639. [CrossRef] [PubMed]

122. Spartz, E.J.; Freeman, G.M., Jr.; Brown, K.; Farhadian, B.; Thienemann, M.; Frankovich, J. Course of neuropsychiatric symptoms after introduction and removal of nonsteroidal anti-inflammatory drugs: A pediatric observational study. *J. Child Adolesc. Psychopharmacol.* **2017**, *27*, 652–659. [CrossRef] [PubMed]

123. Murphy, T.K.; Brennan, E.M.; Johnco, C.; Parker-Athill, E.C.; Miladinovic, B.; Storch, E.A.; Lewin, A.B. A double-blind randomized placebo-controlled pilot study of azithromycin in youth with acute-onset obsessive-compulsive disorder. *J. Child Adolesc. Psychopharmacol.* **2017**, *27*, 640–651. [CrossRef] [PubMed]

124. Williams, K.A.; Swedo, S.E.; Farmer, C.A.; Grantz, H.; Grant, P.J.; D'Souza, P.; Hommer, R.; Katsovich, L.; King, R.A.; Leckman, J.F. Randomized, controlled trial of intravenous immunoglobulin for pediatric autoimmune neuropsychiatric disorders associated with streptococcal infections. *J. Am. Acad. Child Adolesc. Psychiatry* **2016**, *55*, 860–867.e2. [CrossRef] [PubMed]

125. Sigra, S.; Hesselmark, E.; Bejerot, S. Treatment of PANDAS and PANS: A systematic review. *Neurosci. Biobehav. Rev.* **2018**, *86*, 51–65. [CrossRef] [PubMed]

126. Brown, K.D.; Farmer, C.; Freeman, G.M., Jr.; Spartz, E.J.; Farhadian, B.; Thienemann, M.; Frankovich, J. Effect of early and prophylactic nonsteroidal anti-inflammatory drugs on flare duration in pediatric acute-onset neuropsychiatric syndrome: An observational study of patients followed by an academic community-based pediatric acute-onset neuropsychiatric syndrome clinic. *J. Child Adolesc. Psychopharmacol.* **2017**, *27*, 619–628. [PubMed]

127. Nadeau, J.M.; Jordan, C.; Selles, R.R.; Wu, M.S.; King, M.A.; Patel, P.D.; Hanks, C.E.; Arnold, E.B.; Lewin, A.B.; Murphy, T.K.; et al. A pilot trial of cognitive-behavioral therapy augmentation of antibiotic treatment in youth with pediatric acute-onset neuropsychiatric syndrome-related obsessive-compulsive disorder. *J. Child Adolesc. Psychopharmacol.* **2015**, *25*, 337–343. [CrossRef] [PubMed]

128. Latimer, M.E.; L'Etoile, N.; Seidlitz, J.; Swedo, S.E. Therapeutic plasma apheresis as a treatment for 35 severely ill children and adolescents with pediatric autoimmune neuropsychiatric disorders associated with streptococcal infections. *J. Child Adolesc. Psychopharmacol.* **2015**, *25*, 70–75. [CrossRef] [PubMed]

129. Demesh, D.; Virbalas, J.M.; Bent, J.P. The role of tonsillectomy in the treatment of pediatric autoimmune neuropsychiatric disorders associated with streptococcal infections (PANDAS). *JAMA Otolaryngol. Head Neck Surg.* **2015**, *141*, 272–275. [CrossRef] [PubMed]

130. Pavone, P.; Rapisarda, V.; Serra, A.; Nicita, F.; Spalice, A.; Parano, E.; Rizzo, R.; Maiolino, L.; Di Mauro, P.; Vitaliti, G.; et al. Pediatric autoimmune neuropsychiatric disorder associated with group a streptococcal infection: The role of surgical treatment. *Int. J. Immunopathol. Pharmacol* **2014**, *27*, 371–378. [CrossRef] [PubMed]

131. Bernstein, G.A.; Victor, A.M.; Pipal, A.J.; Williams, K.A. Comparison of clinical characteristics of pediatric autoimmune neuropsychiatric disorders associated with streptococcal infections and childhood obsessive-compulsive disorder. *J. Child Adolesc. Psychopharmacol.* **2010**, *20*, 333–340. [CrossRef] [PubMed]

132. Storch, E.A.; Murphy, T.K.; Geffken, G.R.; Mann, G.; Adkins, J.; Merlo, L.J.; Duke, D.; Munson, M.; Swaine, Z.; Goodman, W.K. Cognitive-behavioral therapy for PANDAS-related obsessive-compulsive disorder: Findings from a preliminary waitlist controlled open trial. *J. Am. Acad. Child Adolesc. Psychiatry* **2006**, *45*, 1171–1178. [CrossRef] [PubMed]

133. Snider, L.A.; Lougee, L.; Slattery, M.; Grant, P.; Swedo, S.E. Antibiotic prophylaxis with azithromycin or penicillin for childhood-onset neuropsychiatric disorders. *Biol. Psychiatry* **2005**, *57*, 788–792. [CrossRef] [PubMed]

134. Garvey, M.A.; Perlmutter, S.J.; Allen, A.J.; Hamburger, S.; Lougee, L.; Leonard, H.L.; Witowski, M.E.; Dubbert, B.; Swedo, S.E. A pilot study of penicillin prophylaxis for neuropsychiatric exacerbations triggered by streptococcal infections. *Biol. Psychiatry* **1999**, *45*, 1564–1571. [CrossRef]

135. Shalbafan, M.; Mohammadinejad, P.; Shariat, S.V.; Alavi, K.; Zeinoddini, A.; Salehi, M.; Askari, N.; Akhondzadeh, S. Celecoxib as an adjuvant to fluvoxamine in moderate to severe obsessive-compulsive disorder: A double-blind, placebo-controlled, randomized trial. *Pharmacopsychiatry* **2015**, *48*, 136–140. [CrossRef] [PubMed]

136. Sayyah, M.; Boostani, H.; Pakseresht, S.; Malayeri, A. A preliminary randomized double-blind clinical trial on the efficacy of celecoxib as an adjunct in the treatment of obsessive-compulsive disorder. *Psychiatry Res.* **2011**, *189*, 403–406. [CrossRef] [PubMed]

137. Esalatmanesh, S.; Abrishami, Z.; Zeinoddini, A.; Rahiminejad, F.; Sadeghi, M.; Najarzadegan, M.R.; Shalbafan, M.R.; Akhondzadeh, S. Minocycline combination therapy with fluvoxamine in moderate-to-severe obsessive-compulsive disorder: A placebo-controlled, double-blind, randomized trial. *Psychiatry Clin. Neurosci.* **2016**, *70*, 517–526. [CrossRef] [PubMed]

138. Rodriguez, C.I.; Bender, J., Jr.; Marcus, S.M.; Snape, M.; Rynn, M.; Simpson, H.B. Minocycline augmentation of pharmacotherapy in obsessive-compulsive disorder: An open-label trial. *J. Clin. Psychiatry* **2010**, *71*, 1247–1249. [CrossRef] [PubMed]

139. Ghanizadeh, A.; Mohammadi, M.R.; Bahraini, S.; Keshavarzi, Z.; Firoozabadi, A.; Alavi Shoshtari, A. Efficacy of N-acetylcysteine augmentation on obsessive compulsive disorder: A multicenter randomized double blind placebo controlled clinical trial. *Iran. J. Psychiatry* **2017**, *12*, 134–141. [PubMed]

140. Costa, D.L.C.; Diniz, J.B.; Requena, G.; Joaquim, M.A.; Pittenger, C.; Bloch, M.H.; Miguel, E.C.; Shavitt, R.G. Randomized, double-blind, placebo-controlled trial of N-acetylcysteine augmentation for treatment-resistant obsessive-compulsive disorder. *J. Clin. Psychiatry* **2017**, *78*, e766–e773. [CrossRef] [PubMed]

141. Paydary, K.; Akamaloo, A.; Ahmadipour, A.; Pishgar, F.; Emamzadehfard, S.; Akhondzadeh, S. N-acetylcysteine augmentation therapy for moderate-to-severe obsessive-compulsive disorder: Randomized, double-blind, placebo-controlled trial. *J. Clin. Pharm. Ther.* **2016**, *41*, 214–219. [CrossRef] [PubMed]

142. Sarris, J.; Oliver, G.; Camfield, D.A.; Dean, O.M.; Dowling, N.; Smith, D.J.; Murphy, J.; Menon, R.; Berk, M.; Blair-West, S.; et al. N-acetyl cysteine (NAC) in the treatment of obsessive-compulsive disorder: A 16-week, double-blind, randomised, placebo-controlled study. *CNS Drugs* **2015**, *29*, 801–809. [CrossRef] [PubMed]

143. Afshar, H.; Roohafza, H.; Mohammad-Beigi, H.; Haghighi, M.; Jahangard, L.; Shokouh, P.; Sadeghi, M.; Hafezian, H. N-acetylcysteine add-on treatment in refractory obsessive-compulsive disorder: A randomized, double-blind, placebo-controlled trial. *J Clin. Psychopharmacol.* **2012**, *32*, 797–803. [CrossRef] [PubMed]

144. Hamdani, N.; Daban-Huard, C.; Lajnef, M.; Richard, J.R.; Delavest, M.; Godin, O.; Le Guen, E.; Vederine, F.E.; Lepine, J.P.; Jamain, S.; et al. Relationship between *Toxoplasma gondii* infection and bipolar disorder in a french sample. *J. Affect. Disord* **2013**, *148*, 444–448. [CrossRef] [PubMed]

145. Fond, G.; Capdevielle, D.; Macgregor, A.; Attal, J.; Larue, A.; Brittner, M.; Ducasse, D.; Boulenger, J.P. *Toxoplasma gondii*: A potential role in the genesis of psychiatric disorders. *Encephale* **2013**, *39*, 38–43. [CrossRef] [PubMed]

146. Strittmatter, C.; Lang, W.; Wiestler, O.D.; Kleihues, P. The changing pattern of human immunodeficiency virus-associated cerebral toxoplasmosis: A study of 46 postmortem cases. *Acta Neuropathol.* **1992**, *83*, 475–481. [CrossRef] [PubMed]

147. Ueno, N.; Lodoen, M.B. From the blood to the brain: Avenues of eukaryotic pathogen dissemination to the central nervous system. *Curr. Opin. Microbiol.* **2015**, *26*, 53–59. [CrossRef] [PubMed]

148. Courret, N.; Darche, S.; Sonigo, P.; Milon, G.; Buzoni-Gatel, D.; Tardieux, I. CD11c- and CD11b-expressing mouse leukocytes transport single *Toxoplasma gondii* tachyzoites to the brain. *Blood* **2006**, *107*, 309–316. [CrossRef] [PubMed]

149. Dellacasa-Lindberg, I.; Fuks, J.M.; Arrighi, R.B.; Lambert, H.; Wallin, R.P.; Chambers, B.J.; Barragan, A. Migratory activation of primary cortical microglia upon infection with *Toxoplasma gondii*. *Infect. Immun.* **2011**, *79*, 3046–3052. [CrossRef] [PubMed]

150. Notarangelo, F.M.; Wilson, E.H.; Horning, K.J.; Thomas, M.A.; Harris, T.H.; Fang, Q.; Hunter, C.A.; Schwarcz, R. Evaluation of kynurenine pathway metabolism in *Toxoplasma gondii*-infected mice: Implications for schizophrenia. *Schizophr. Res.* **2014**, *152*, 261–267. [CrossRef] [PubMed]

151. Mosienko, V.; Beis, D.; Pasqualetti, M.; Waider, J.; Matthes, S.; Qadri, F.; Bader, M.; Alenina, N. Life without brain serotonin: Reevaluation of serotonin function with mice deficient in brain serotonin synthesis. *Behav. Brain Res.* **2015**, *277*, 78–88. [CrossRef] [PubMed]

152. Mandi, Y.; Vecsei, L. The kynurenine system and immunoregulation. *J. Neural. Transm.* **2012**, *119*, 197–209. [CrossRef] [PubMed]

153. Yadav, M.C.; Burudi, E.M.; Alirezaei, M.; Flynn, C.C.; Watry, D.D.; Lanigan, C.M.; Fox, H.S. IFN-gamma-induced ido and wrs expression in microglia is differentially regulated by IL-4. *Glia* **2007**, *55*, 1385–1396. [CrossRef] [PubMed]

154. Guillemin, G.J.; Smythe, G.; Takikawa, O.; Brew, B.J. Expression of indoleamine 2,3-dioxygenase and production of quinolinic acid by human microglia, astrocytes, and neurons. *Glia* **2005**, *49*, 15–23. [CrossRef] [PubMed]

155. Daubener, W.; MacKenzie, C.R. IFN-gamma activated indoleamine 2,3-dioxygenase activity in human cells is an antiparasitic and an antibacterial effector mechanism. *Adv. Exp. Med. Biol.* **1999**, *467*, 517–524. [PubMed]

156. Booij, L.; Van der Does, W.; Benkelfat, C.; Bremner, J.D.; Cowen, P.J.; Fava, M.; Gillin, C.; Leyton, M.; Moore, P.; Smith, K.A.; et al. Predictors of mood response to acute tryptophan depletion: A reanalysis. *Neuropsychopharmacology* **2002**, *27*, 852–861. [CrossRef]

157. Romanelli, R.J.; Wu, F.M.; Gamba, R.; Mojtabai, R.; Segal, J.B. Behavioral therapy and serotonin reuptake inhibitor pharmacotherapy in the treatment of obsessive-compulsive disorder: A systematic review and meta-analysis of head-to-head randomized controlled trials. *Depress. Anxiety* **2014**, *31*, 641–652. [CrossRef] [PubMed]

158. Yatham, L.N.; Liddle, P.F.; Sossi, V.; Erez, J.; Vafai, N.; Lam, R.W.; Blinder, S. Positron emission tomography study of the effects of tryptophan depletion on brain serotonin(2) receptors in subjects recently remitted from major depression. *Arch. Gen. Psychiatry* **2012**, *69*, 601–609. [CrossRef] [PubMed]

159. Hughes, M.M.; Carballedo, A.; McLoughlin, D.M.; Amico, F.; Harkin, A.; Frodl, T.; Connor, T.J. Tryptophan depletion in depressed patients occurs independent of kynurenine pathway activation. *Brain Behav. Immun.* **2012**, *26*, 979–987. [CrossRef] [PubMed]

160. Prandovszky, E.; Gaskell, E.; Martin, H.; Dubey, J.P.; Webster, J.P.; McConkey, G.A. The neurotropic parasite *Toxoplasma gondii* increases dopamine metabolism. *PLoS ONE* **2011**, *6*, e23866. [CrossRef] [PubMed]

161. Soares-Cunha, C.; Coimbra, B.; Sousa, N.; Rodrigues, A.J. Reappraising striatal D1- and D2-neurons in reward and aversion. *Neurosci. Biobehav. Rev.* **2016**, *68*, 370–386. [CrossRef] [PubMed]

162. Wichers, M.C.; Koek, G.H.; Robaeys, G.; Verkerk, R.; Scharpe, S.; Maes, M. Ido and interferon-alpha-induced depressive symptoms: A shift in hypothesis from tryptophan depletion to neurotoxicity. *Mol. Psychiatry* **2005**, *10*, 538–544. [CrossRef] [PubMed]

163. Prendergast, G.C.; Malachowski, W.J.; Mondal, A.; Scherle, P.; Muller, A.J. Indoleamine 2,3-dioxygenase and its therapeutic inhibition in cancer. *Int. Rev. Cell Mol. Biol.* **2018**, *336*, 175–203. [PubMed]

164. Reus, G.Z.; Becker, I.R.T.; Scaini, G.; Petronilho, F.; Oses, J.P.; Kaddurah-Daouk, R.; Ceretta, L.B.; Zugno, A.I.; Dal-Pizzol, F.; Quevedo, J.; et al. The inhibition of the kynurenine pathway prevents behavioral disturbances and oxidative stress in the brain of adult rats subjected to an animal model of schizophrenia. *Prog. Neuropsychopharmacol. Biol. Psychiatry* **2018**, *81*, 55–63. [CrossRef] [PubMed]

165. Fineberg, N.A.; Reghunandanan, S.; Simpson, H.B.; Phillips, K.A.; Richter, M.A.; Matthews, K.; Stein, D.J.; Sareen, J.; Brown, A.; Sookman, D.; et al. Obsessive-compulsive disorder (OCD): Practical strategies for pharmacological and somatic treatment in adults. *Psychiatry Res.* **2015**, *227*, 114–125. [CrossRef] [PubMed]

166. Perlmutter, S.J.; Leitman, S.F.; Garvey, M.A.; Hamburger, S.; Feldman, E.; Leonard, H.L.; Swedo, S.E. Therapeutic plasma exchange and intravenous immunoglobulin for obsessive-compulsive disorder and tic disorders in childhood. *Lancet* **1999**, *354*, 1153–1158. [CrossRef]

167. Cooperstock, M.S.; Swedo, S.E.; Pasternack, M.S.; Murphy, T.K.; PANS/PANDAS Consortium. Clinical management of pediatric acute-onset neuropsychiatric syndrome: Part III—Treatment and prevention of infections. *J. Child Adolesc. Psychopharmacol.* **2017**, *27*, 594–606. [CrossRef]

168. National Institute of Mental Health (NIMH). Information about PANS/PANDAS. Available online: https://www.nimh.nih.gov/labs-at-nimh/research-areas/clinics-and-labs/sbp/information-about-pans-pandas.shtml (accessed on 30 March 2018).

169. Frankovich, J.; Swedo, S.; Murphy, T.; Dale, R.C.; Agalliu, D.; Williams, K.; Daines, M.; Hornig, M.; Chugani, H.; Sanger, T.; et al. Clinical Management of pediatric acute-onset neuropsychiatric syndrome: Part II—Use of immunomodulatory therapies. *J. Child Adolesc. Psychopharmacol.* **2017**, *27*, 574–593. [CrossRef]

170. Kellner, M.; Nowack, S.; Wortmann, V.; Yassouridis, A.; Wiedemann, K. Does pregnenolone enhance exposure therapy in obsessive-compulsive disorder?—A pilot, interim report of a randomized, placebo-controlled, double-blind study. *Pharmacopsychiatry* **2016**, *49*, 79–81. [CrossRef] [PubMed]

171. Murphy, T.K.; Parker-Athill, E.C.; Lewin, A.B.; Storch, E.A.; Mutch, P.J. Cefdinir for recent-onset pediatric neuropsychiatric disorders: A pilot randomized trial. *J. Child Adolesc. Psychopharmacol.* **2015**, *25*, 57–64. [CrossRef] [PubMed]

172. Kalueff, A.V.; Stewart, A.M.; Song, C.; Berridge, K.C.; Graybiel, A.M.; Fentress, J.C. Neurobiology of rodent self-grooming and its value for translational neuroscience. *Nat. Rev. Neurosci.* **2016**, *17*, 45–59. [CrossRef] [PubMed]

173. Welch, J.M.; Lu, J.; Rodriguiz, R.M.; Trotta, N.C.; Peca, J.; Ding, J.D.; Feliciano, C.; Chen, M.; Adams, J.P.; Luo, J.; et al. Cortico-striatal synaptic defects and OCD-like behaviours in *Sapap3*-mutant mice. *Nature* **2007**, *448*, 894–900. [CrossRef] [PubMed]

174. Gehring, W.J.; Hiromi, Y. Homeotic genes and the homeobox. *Annu. Rev. Genet.* **1986**, *20*, 147–173. [CrossRef] [PubMed]

175. Chen, S.K.; Tvrdik, P.; Peden, E.; Cho, S.; Wu, S.; Spangrude, G.; Capecchi, M.R. Hematopoietic origin of pathological grooming in *Hoxb8* mutant mice. *Cell* **2010**, *141*, 775–785. [CrossRef] [PubMed]

176. Brimberg, L.; Benhar, I.; Mascaro-Blanco, A.; Alvarez, K.; Lotan, D.; Winter, C.; Klein, J.; Moses, A.E.; Somnier, F.E.; Leckman, J.F.; et al. Behavioral, pharmacological, and immunological abnormalities after streptococcal exposure: A novel rat model of sydenham chorea and related neuropsychiatric disorders. *Neuropsychopharmacology* **2012**, *37*, 2076–2087. [CrossRef] [PubMed]

177. Lotan, D.; Cunningham, M.; Joel, D. Antibiotic treatment attenuates behavioral and neurochemical changes induced by exposure of rats to group a streptococcal antigen. *PLoS ONE* **2014**, *9*, e101257. [CrossRef] [PubMed]

178. Pauls, D.L.; Raymond, C.L.; Stevenson, J.M.; Leckman, J.F. A family study of gilles de la tourette syndrome. *Am. J. Hum. Genet.* **1991**, *48*, 154–163. [PubMed]

179. Eapen, V.; Pauls, D.L.; Robertson, M.M. Evidence for autosomal dominant transmission in tourette's syndrome. United kingdom cohort study. *Br. J. Psychiatry* **1993**, *162*, 593–596. [CrossRef] [PubMed]

180. Kataoka, Y.; Kalanithi, P.S.; Grantz, H.; Schwartz, M.L.; Saper, C.; Leckman, J.F.; Vaccarino, F.M. Decreased number of parvalbumin and cholinergic interneurons in the striatum of individuals with tourette syndrome. *J. Comp. Neurol.* **2010**, *518*, 277–291. [CrossRef] [PubMed]

181. Frick, L.R.; Williams, K.; Pittenger, C. Microglial dysregulation in psychiatric disease. *Clin. Dev. Immunol.* **2013**, *2013*, 608654. [CrossRef] [PubMed]

182. Morer, A.; Chae, W.; Henegariu, O.; Bothwell, A.L.; Leckman, J.F.; Kawikova, I. Elevated expression of MCP-1, IL-2 and PTPR-N in basal ganglia of tourette syndrome cases. *Brain Behav. Immun.* **2010**, *24*, 1069–1073. [CrossRef] [PubMed]

183. Bilbo, S.D.; Schwarz, J.M. The immune system and developmental programming of brain and behavior. *Front. Neuroendocrinol.* **2012**, *33*, 267–286. [CrossRef] [PubMed]

184. Attwells, S.; Setiawan, E.; Wilson, A.A.; Rusjan, P.M.; Mizrahi, R.; Miler, L.; Xu, C.; Richter, M.A.; Kahn, A.; Kish, S.J.; et al. Inflammation in the neurocircuitry of obsessive-compulsive disorder. *JAMA Psychiatry* **2017**, *74*, 833–840. [CrossRef] [PubMed]

185. Scott, G.; Zetterberg, H.; Jolly, A.; Cole, J.H.; De Simoni, S.; Jenkins, P.O.; Feeney, C.; Owen, D.R.; Lingford-Hughes, A.; Howes, O.; et al. Minocycline reduces chronic microglial activation after brain trauma but increases neurodegeneration. *Brain* **2018**, *141*, 459–471. [CrossRef] [PubMed]

186. Fernandez, T.V.; Leckman, J.F.; Pittenger, C. Genetic susceptibility in obsessive-compulsive disorder. *Handb. Clin. Neurol.* **2018**, *148*, 767–781. [PubMed]

187. Pishva, E.; Drukker, M.; Viechtbauer, W.; Decoster, J.; Collip, D.; van Winkel, R.; Wichers, M.; Jacobs, N.; Thiery, E.; Derom, C.; et al. Epigenetic genes and emotional reactivity to daily life events: A multi-step gene-environment interaction study. *PLoS ONE* **2014**, *9*, e100935. [CrossRef] [PubMed]

188. Carlsson, E.; Frostell, A.; Ludvigsson, J.; Faresjo, M. Psychological stress in children may alter the immune response. *J. Immunol.* **2014**, *192*, 2071–2081. [CrossRef] [PubMed]

189. Leboyer, M.; Oliveira, J.; Tamouza, R.; Groc, L. Is it time for immunopsychiatry in psychotic disorders? *Psychopharmacology* **2016**, *233*, 1651–1660. [CrossRef] [PubMed]

190. Leboyer, M.; Berk, M.; Yolken, R.H.; Tamouza, R.; Kupfer, D.; Groc, L. Immuno-psychiatry: An agenda for clinical practice and innovative research. *BMC Med.* **2016**, *14*, 173. [CrossRef] [PubMed]

MDPI

St. Alban-Anlage 66

4052 Basel

Switzerland

Tel. +41 61 683 77 34

Fax +41 61 302 89 18

www.mdpi.com

Brain Sciences Editorial Office

E-mail: brainsci@mdpi.com

www.mdpi.com/journal/brainsci

www.ingramcontent.com/pod-product-compliance
Lightning Source LLC
Chambersburg PA
CBHW051916210326
41597CB00033B/6159